NOW 2 KNOW!

Calculus 2

by T. G. D'Alberto

Pithy Professor Publishing Company

Brighton, CO

Published by

Pithy Professor Publishing Company, LLC
PO Box 33824
Northglenn, CO 80233

Copyright © 2014 by
Pithy Professor Publishing Company, LLC

ISBN: 978-09882054-3-7

Library of Congress Control Number: 2014901634

Printed in the United States of America

About the Author

Dr. Tiffanie G. D'Alberto has a Ph.D. in Electrical & Computer Engineering from Cornell University and a B.S. and M.S. in Electrical Engineering from Virginia Polytechnic Institute & State University.

She has worked for over a decade in the telecommunications and aerospace industries as a scientist, program manager, and supervisor. She has engaged in numerous opportunities for tutoring, teaching, and mentoring throughout her career and schooling.

In her spare time, Tiffanie enjoys oil painting, drawing, reading, sewing, and running. She's a huge fan of Star Trek, Renaissance Festivals, and animals.

Tiffanie lives in St. Croix with her fiancé, Colin, and their many wonderful pets.

Dedication

To my dearest Colin, who inspires me, encourages me, and supports me. I could never thank you enough.

To my high school Calculus teacher, Mr. Klima, who showed me how easy Calculus can be.

Acknowledgements

I always thank my family first: My parents for the foundation, the push, and the belief in me all along; My fiancé for his inspiration, encouragement, and unending support.

A huge thanks goes to my many excellent math teachers from middle school to high school to college that not only taught the material but also taught the way of thinking necessary to excel in these subjects.

I'd also like to acknowledge **Calculus with Analytical Geometry**, Second Alternate Edition by Earl W. Swokowski. His clear, accurate, and thorough discussion of Calculus has aided in my understanding of the subject and in the development of this book. Any good book needs good graphics, and Softonic.com provided wonderful 3-D renditions of multi-variable functions.

Finally, I'd like to thank Amazon.com for their excellent publish-on-demand service that enables books such as these, and you, the reader, for making this investment in your future.

Table of Contents

Introduction

Welcome!

I've taken quite a number of math-intensive technical courses - most went very well, but a few went very poorly. Through many years of schooling, tutoring, and teaching, I've learned what does and does not work in terms of learning math. This book is shaped after those lessons.

The philosophy of this book is three-fold:
1. **To excel at math is to understand math.** It's NOT memorization! Think of it as any game you would play. If you *understand* the rules and how to use them to your advantage, you can play the game well. That's how you should learn math.
2. **To understand math, you need the story** . The story is the logic flow that allows you to keep building on your *understanding*.
3. **To understand math, you also need the big picture.** The big picture is the outline of the logic, or story. Once you see the big picture, it's easier for you to put the details of the story into their proper places.

This course is comprised of two parts: multi-variable calculus and infinite sequences and series. In the first part, you find that Calculus 1, 2, 3, 4... are just variations on a theme. You'll also find that bookkeeping becomes more important.

The second part, infinite sequences and series, covers topics that are difficult to many students. The material may seem irrelevant and disconnected at first, but it can be very important in future course- and career-work. Just take it slowly and study the examples.

Be open to changing the way you think. Once you get how to learn this subject, you'll get the A's. I wish you great luck!

Layout:

The layout of this text is different from most academic books:

1. **The problem sets are saved to the end of the book.** While it is true that you should master one chapter before continuing to the next, and mastery takes practice, placing problem sets between chapters tends to take away from the flow of the story. In this book, you can read from beginning to end to understand the logical progression of the course, or stop to do problem sets as you desire.

2. **Solution sets give the critical steps to get the answers, not just the answers.** Because this is not a textbook for a classroom, there is no need to keep the "secret sauce" from you. Use the problem sets as drills or as further examples for study.

3. **Appendix A is an overall summary of the entire book.** It helps you visualize the big picture and logic flow to give you a framework into which you can organize the details.

In addition, the following visual markers will help you navigate the material...

Key terms defined for the first time are **bolded** and also found in the index.

Important equations are shown as:

> ***important equations***

Illustrative graphics and additional notes are shown on the side to accompany the text.

Finally, examples are given as supplements to the text as well as for illustration:

> *Example* ▷ This is an example to illustrate a point or to give further definition. Skip it if you feel very comfortable with the material presented thus far.

Notation

The following notation is often helpful when talking about Math. Some of this notation will be used in the text.

Closed interval – includes points a and b: [a,b]

Open interval – up to but not including points a and b: (a,b)

$\exists$	There exists
Я or Я	With respect to
$\forall$	For all; For each; For every
$\ni$	Such that
$\in$	Is an element of
\	Except
iff or $\Leftrightarrow$	If and only if
b/c	Because
b/w	Between
f^n	Function
⊙⊙⊙	On the order of (think 3 O's)
$\rightarrow$	Implies, Is given by
$\therefore$	Therefore
DNE	Do/does not exist
α	Proportional to
$\sim, \approx, \cong$	Equivalent to; Approximately equal to
$\equiv$	Is assigned to; Is defined as; Is forced to equal to
$\neq$	Is not equal to
//	Parallel to
$\perp$	Perpendicular to; Orthogonal to

Final Note: Variables are usually u, v, w, x, y, z, and constants are usually a, b, c, k, and n

Part 1: Multi-Variable Calculus

Chapter 1: Functions of Multiple Variables

Functions:

A single-variable function, like $y = f(x)$, has only one variable in the operand (in parentheses). The x is the **independent variable** and its allowable values make up the **domain** of the function. The y is the **dependent variable** (on x), and its values make up the **range** of the function.

In order for $f(x)$ to be a function, it must pass the vertical line test. That is, for each x value, there is at most one y value. When $f(x)$ is plotted, a scanning vertical line will never intersect more than one point at a time.

All of the above is true for a **multi-variable function**, as well. As an example, consider the two variable function:

$$w = f(x, y) = 4x^2 + 2y^2.$$

In this case, x and y are the **independent variables** that make up the **domain**, and w is the **range** of $f(x, y)$ and the **dependent variable** defined by x and y.

You can think of the x-y plane and plotting the function along a third axis to give a three dimensional graph like that shown to the right. In order to be a function, the following must be met for every x-y pair:

> **If w is a function, then**
> **for any point in the domain,**
> **there can only exist at most one range value.**

In other words, for any given x-y pair, there is at most one value for w. It's like doing the vertical line test with the line parallel to the w axis (or perpendicular to the x-y plane). The same is true for functions of more than two variables, but the picture is harder to imagine.

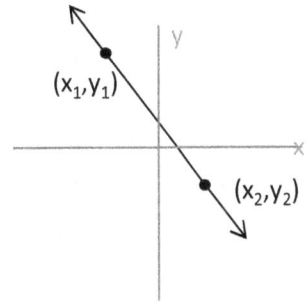

A line defined by two points: (x_1, y_1) and (x_2, y_2).

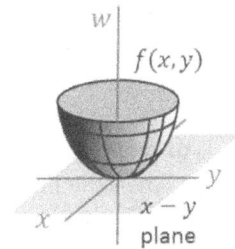

A two variable function.

The following are examples of functions of two variables:

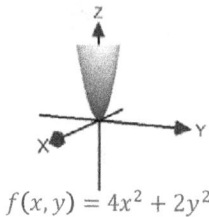

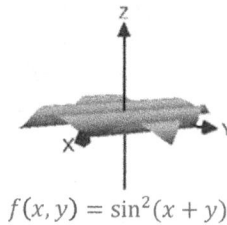

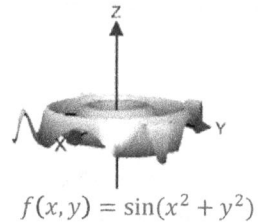

$$f(x,y) = 4x^2 + 2y^2 \qquad f(x,y) = \sin^2(x+y) \qquad f(x,y) = \sin(x^2 + y^2)$$

Continuity:

Just as for single-variable calculus, it's important to know where a function satisfies **continuity**, or is **continuous**. Most of the time that's easy because you can solve for a divide by zero situation or the definition of the function is split between two domains. But, there is a formal definition based on the idea of limits which is called the **Two Path Rule**:

> If the limit of a function at a point P
> is different when approaching P from different directions,
> then the limit of the function at P does not exist,
> and the function is discontinuous at P.

$$f(x,y) = \frac{\sin(x^2+y^2)}{y}$$

Example of discontinuities along x-axis where $y = 0$. Approaching from $+y$ gives $+\infty$, from $-y$ gives $-\infty$

For a two variable function, we can write the above definition in symbolic form:

> If $\lim\limits_{(x,y)\to(a,b)^A} f(x,y) \neq \lim\limits_{(x,y)\to(a,b)^B} f(x,y)$,
>
> Then $\exists$ a discontinuity at (a, b).

where A indicates approaching from direction A, and B indicates approaching from direction B.

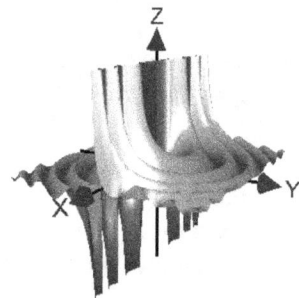

Example

Use the two path rule to show the following function is not continuous:

$$f(x,y) = \begin{cases} x^2 + y^2; & y > 1 \\ x + y; & y < 1 \end{cases}$$

This is an example where the domain is broken into two parts. Along the line $y = 1$, the function is undefined:

$$\lim_{y \to 1^+} x^2 + y^2 = x^2 + 1$$

$$\lim_{y \to 1^-} x + y = x + 1$$

Approaching the discontinuity from different directions gives different answers for the limit.

Example

Use the two path rule to show the following function is not continuous:

$$f(x,y) = \frac{x^2 + 2y^2}{x^2 - y^2}$$

$$f(x,y) = \frac{x^2 + 2y^2}{x^2 - y^2}$$

By inspection, the function is undefined at the point (0,0), so let's approach that point along a line $(y = mx + b)$ that passes through $(x,y) = (0,0)$:

Let $y = mx + 0$;

$$\lim_{(x,y) \to (0,0)} \frac{x^2 + 2y^2}{x^2 - y^2} = \lim_{x \to 0} \frac{x^2 + 2(mx)^2}{x^2 - (mx)^2}$$

$$= \lim_{x \to 0} \frac{x^2(1 + 2m^2)}{x^2(1 - m^2)} = \frac{(1 + 2m^2)}{(1 - m^2)}$$

$y = m_1 x$

$y = m_2 x$

So, if $m = 1$, the limit is ∞; if $m = 2$, the limit is -3; if $m = 3$, the limit is -19/8. The limit yields different answers depending on which direction the point (0,0) is approached.

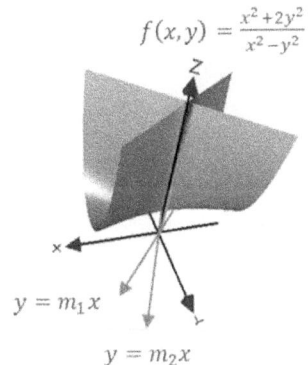

Discontinuity at $(x,y) = (0,0)$ gives different answers for limit when approaching along different $y = mx$ lines.

Chapter 2: Partial Derivatives

Formal Definition:

Just like with single variable calculus, you can take a derivative of a multi-variable function. You just have to pick which independent variable you want the derivative to address. In the example $w = f(x, y)$, you need to decide whether to take the derivative with respect to x or to y.

Since you do a derivative with respect to only one independent variable at a time, these derivatives are called **partial derivatives**. The notation is slightly different than the single-variable case to indicate the *partial* derivative and which variable has been chosen:

Derivative of $w = f(x)$	$f'(x)$; $\frac{d}{dx}f(x)$; $D_x[f(x)]$; w'; $\frac{dw}{dx}$; dw; $D_x w$
Partial Derivative of $w = f(x, y) \ \text{Ɓ} \ x$	$f_x(x, y)$; $\frac{\partial}{\partial x}f(x, y)$; $\frac{\partial f}{\partial x}$; f_x; w_x; $\frac{\partial w}{\partial x}$

And, just like with single variable calculus, derivatives with multiple variables have a formal definition based on limits. As you draw a line between two points on a function and allow those points to get closer and closer in one dimension, you get the slope of the function in that dimension near those points. You can also generalize to find an overall expression for the slope everywhere (in the allowable domain). The generalization is called the derivative.

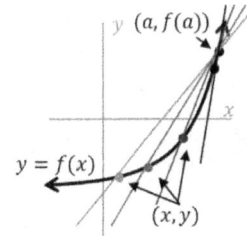

Recall from Calculus 1: As point $(x, y) \to (a, f(a))$, the line between the two points becomes the tangent of the curve at point $(a, f(a))$.

Considering again a two variable function, $f(x, y)$, we can write the partial derivative with respect to x and y as follows:

$$f_x(x, y) = \lim_{h \to 0} \frac{f(x+h, y) - f(x, y)}{h};$$

$$f_y(x, y) = \lim_{h \to 0} \frac{f(x, y+h) - f(x, y)}{h}.$$

First Partial Derivative:

Taking **first partial derivatives** is similar to taking first full derivatives. The only difference is that when you take the derivative with respect to one variable, you treat all other variables as constants.

Example ▷ Find $\frac{\partial f}{\partial x}$ and $\frac{\partial f}{\partial y}$:

$$f(x,y) = x^2 y + y^3 x$$

For $\frac{\partial f}{\partial x}$, we treat y as a constant:

$$\frac{\partial f}{\partial x} = 2xy + y^3$$

For $\frac{\partial f}{\partial y}$, we treat x as a constant:

$$\frac{\partial f}{\partial y} = x^2 + 3xy^2$$

Example ▷ Find $\frac{\partial f}{\partial x}$ and $\frac{\partial f}{\partial y}$:

$$f(x,y) = (x + y)(x^2 + xy + 4y^2)$$

For $\frac{\partial f}{\partial x}$, we treat y as a constant:

$$\frac{\partial f}{\partial x} = (x + y)(2x + y) + (x^2 + xy + 4y^2)$$

For $\frac{\partial f}{\partial y}$, we treat x as a constant:

$$\frac{\partial f}{\partial y} = (x + y)(x + 8y) + (x^2 + xy + 4y^2)$$

Second Partial Derivative:

The **second** (and **higher order**) **partial derivatives** are done the same as the first partial derivative. No matter which variable you choose in your first partial derivative, you may choose any independent variable in the second partial derivative. The notation shows which variables and in which order the derivatives are taken.

Again consider the example, $w = f(x, y)$. There are four possible combinations of second partial derivatives. In the table below, the notation 1st ℟ x means the first derivative with respect to x:

	2nd ℟ x	2nd ℟ y
1st ℟ x	$f_{xx}, (f_x)_x, \frac{\partial}{\partial x}f_x, \frac{\partial}{\partial x}\left(\frac{\partial f}{\partial x}\right), \frac{\partial^2 f}{\partial x^2}$	$f_{xy}, (f_x)_y, \frac{\partial}{\partial y}f_x, \frac{\partial}{\partial y}\left(\frac{\partial f}{\partial x}\right), \frac{\partial^2 f}{\partial y \partial x}$
1st ℟ y	$f_{yx}, (f_y)_x, \frac{\partial}{\partial x}f_y, \frac{\partial}{\partial x}\left(\frac{\partial f}{\partial y}\right), \frac{\partial^2 f}{\partial x \partial y}$	$f_{yy}, (f_y)_y, \frac{\partial}{\partial y}f_y, \frac{\partial}{\partial y}\left(\frac{\partial f}{\partial y}\right), \frac{\partial^2 f}{\partial y^2}$

Note that in the notation using subscripts, the order of differentiation starts with the variable closest to the f and works out (left to right). In the notation using the partial derivative symbol, ∂, the order reads from right to left.

Higher order derivatives are formed in the same manner. Take the example, $w = f(x, y, z)$. The 3rd partial derivative with respect to x, then y, then z is:

$$f_{xyz}, ((f_x)_y)_z, \frac{\partial^3 f}{\partial z \partial y \partial x}, \dots$$

One final thing to note is a theorem that states :

> If f_{xx}, f_{xy}, f_{yx}, and f_{yy} are continuous,
>
> then $f_{xy} = f_{yx}$.

13

Example

Find the second derivatives:
$$f(x, y) = x^2 y + y^3 x$$

From a prior example, we know:

$$\frac{\partial f}{\partial x} = 2xy + y^3; \quad \frac{\partial f}{\partial y} = x^2 + 3xy^2$$

Using these results:

$$\frac{\partial^2 f}{\partial x^2} = 2y; \qquad\qquad \frac{\partial^2 f}{\partial y \partial x} = 2x + 3y^2;$$

$$\frac{\partial^2 f}{\partial x \partial y} = 2x + 3y^2; \qquad \frac{\partial^2 f}{\partial y^2} = 6xy.$$

Example

Find the second derivatives:
$$f(x, y) = (x + y)(x^2 + xy + 4y^2)$$

From a prior example, we know:

$$\frac{\partial f}{\partial x} = (x + y)(2x + y) + (x^2 + xy + 4y^2)$$

$$\frac{\partial f}{\partial y} = (x + y)(x + 8y) + (x^2 + xy + 4y^2)$$

Using these results:

$$\frac{\partial^2 f}{\partial x^2} = 2(x + y) + (2x + y) + 2x + y = 6x + 4y;$$

$$\frac{\partial^2 f}{\partial y \partial x} = (x + y) + (2x + y) + 8y + x = 4x + 10y;$$

$$\frac{\partial^2 f}{\partial x \partial y} = (x + y) + (x + 8y) + 2x + y = 4x + 10y;$$

$$\frac{\partial^2 f}{\partial y^2} = 8(x + y) + (x + 8y) + 8y + x = 10x + 24y.$$

Chapter 3: Properties & Rules

Properties of Partial Derivatives:

The properties of partial derivatives are the same as for full derivatives keeping in mind that all other independent variables are treated as constants. Though we've already used these properties in the examples of Chapter 2, they are summarized here for convenience.

Derivatives of Constants are 0

$$\frac{\partial}{\partial x}[c] = 0$$

Power Rule

$$\frac{\partial}{\partial x}[x^n] = nx^{n-1}$$

Pull Out Constants

$$\frac{\partial}{\partial x}[cf(x)] = cf_x$$

Distribution

$$\frac{\partial}{\partial x}[f(x) \pm g(x)] = f_x \pm g_x$$

Product Rule

$$\frac{\partial}{\partial x}[f(x) \cdot g(x)] = fg_x + gf_x$$

Quotient Rule

$$\frac{\partial}{\partial x}\left[\frac{f(x)}{g(x)}\right] = \frac{gf_x - fg_x}{g^2}$$

Increments:

So far, we can talk about slopes in one direction or another with respect to individual independent variables. However, if we want to find the absolute change, or **increment**, in $w = f(x, y)$ when both x and y change by some small amount Δx and Δy, respectively, we use the following:

$$\Delta w = f(x + \Delta x, y + \Delta y) - f(x, y)$$

As the incremental change in x and y becomes small, we may use the following expression instead:

As $\Delta x \to dx$ and $\Delta y \to dy$:

$$dw = \frac{\partial w}{\partial x} dx + \frac{\partial w}{\partial y} dy.$$

Example

Determine how much the function w changes near the point $(2,2)$ using both definitions above for:

$$w = x^2 + y; \quad \Delta x = 0.1; \quad \Delta y = 0.2;$$

Using the first definition:
$$\Delta w = (x + \Delta x)^2 + y + \Delta y - (x^2 + y)$$
$$= 2x\Delta x + \Delta x^2 + \Delta y$$
$$= 4(0.1) + 0.1^2 + 0.2 = 0.61$$

Using the second definition:
$$dx = 0.1; \quad dy = 0.2;$$
$$dw = 2x\,dx + dy = 4(0.1) + 0.2 = 0.6$$

To check, the value of w goes from:
$$2^2 + 2 = 6 \text{ to } 2.1^2 + 2.2 = 6.61,$$
a difference of 0.61.

Had dx and dy been smaller, the second definition would have yielded closer results.

The Chain Rule:

Recall that for full derivatives, the Chain Rule is:

If $w = f(u)$ and $u = g(x)$; i.e. $w = f(g(x))$;

$$\text{then } \frac{dw}{dx} = \frac{dw}{du} \cdot \frac{du}{dx} = f'(u)du$$

In other words, take the derivative of the outside times the derivative of the inside.

The **Chain Rule** can be expanded to the following:

Let $w = f(u, v)$; $u = g(x, y)$; $v = k(x, y)$;
i.e. $w = f(g(x, y), k(x, y))$;

$$\frac{\partial w}{\partial x} = \frac{\partial w}{\partial u} \cdot \frac{\partial u}{\partial x} + \frac{\partial w}{\partial v} \cdot \frac{\partial v}{\partial x} = f_u \cdot u_x + f_v \cdot v_x$$

Example

Find $\frac{\partial w}{\partial x}$ for: $u = xy$; $v = x + y$; $w = u^2 + v^2$;

$$\frac{\partial u}{\partial x} = y; \quad \frac{\partial w}{\partial u} = 2u = 2xy;$$

$$\frac{\partial v}{\partial x} = 1; \quad \frac{\partial w}{\partial v} = 2v = 2x + 2y;$$

$$\frac{\partial w}{\partial u} \cdot \frac{\partial u}{\partial x} + \frac{\partial w}{\partial v} \cdot \frac{\partial v}{\partial x} = 2xy^2 + 2x + 2y$$

Compare this result to directly taking the partial derivative of w with respect to x:

$$w = u^2 + v^2 = x^2y^2 + (x + y)^2$$

$$\frac{\partial w}{\partial x} = 2xy^2 + 2(x + y)$$

The Implicit Chain Rule:

Suppose that instead of having a free-standing function, we have an equation such as $f(x, y) = 0$. Many times the equation can be solved to obtain a form that looks like $y = g(x)$. If that is true, then the equation $f(x, y) = 0$ is said to **implicitly determine** y.

If we wish to find y' for a given equation, we can solve $f(x, y) = 0$ for $y = g(x)$ and then take the derivative. It may be easier, though, to use what I'm calling the **Implicit Chain Rule**:

> If $y = g(x)$ **can be implicitly determined from the equation** $f(x, y) = 0$;
>
> $$\frac{dy}{dx} = -\frac{\partial f/\partial x}{\partial f/\partial y} = \frac{-f_x}{f_y}$$

Example

Assume the following equation can be solved for y and find y':
$$x^3 - 4yx^2 = 2$$

$$f(x, y) = x^3 - 4yx^2 - 2 = 0$$

$$\frac{\partial f}{\partial x} = 3x^2 - 8xy; \quad \frac{\partial f}{\partial y} = -4x^2;$$

$$y' = -\frac{\partial f/\partial x}{\partial f/\partial y} = \frac{8xy - 3x^2}{-4x^2}$$

Compare to:

Solve $f(x, y) = 0$ and get: $y = \frac{2 - x^3}{-4x^2}$;

$$y' = \frac{-4x^2(-3x^2) - (2 - x^2)(-8x)}{(-4x^2)^2} = \frac{-3x^2 + 8xy}{-4x^2}$$

Chapter 4: Max & Min

Critical Points:

In Calculus 1, we found the **critical points** of a function by taking the first derivative, setting it equal to 0, then solving for x. The resulting x values were plugged into the original equation to find the corresponding y values. The case is very similar when finding critical points for multi-variable functions. Consider $w = f(x, y)$:

> **The point (a, b) is a critical point if ONE of the following is true:**
>
> $$\frac{\partial}{\partial x} f(a, b) = \frac{\partial}{\partial y} f(a, b) = 0;$$
>
> $\frac{\partial}{\partial x} f(a, b)$ **does not exist;**
>
> $\frac{\partial}{\partial y} f(a, b)$ **does not exist;**

Test for Max & Min:

In Calculus 1, the test for **maxima** or **minima** was to take the second derivative, plug in the x values of the critical points, and determine if the results were > 0 (minimum), < 0 (maximum), or $= 0$ (inflection). For $w = f(x, y)$, there are two quantities to check.

If $f_x(a, b) = f_y(a, b) = 0$, then (a, b) is a maximum or minimum according to:

	Max	Min	Neither
$f_{xx}(a, b)$	< 0	> 0	
$f_{xx}(a, b) \cdot f_{yy}(a, b) - (f_{xy}(a, b))^2$	> 0	> 0	< 0

Example Find the critical points and determine if they are maxima or minima:

$$x^2 + 2xy + 2y^2$$

Take the first derivatives and set them equal to 0 to find the critical points:

$$\frac{\partial f}{\partial x} = 2x + 2y \equiv 0 \quad \rightarrow \quad x = -y;$$

$$\frac{\partial f}{\partial y} = 2x + 4y \equiv 0 \quad \rightarrow \quad x = -2y;$$

The above equations can only both be true when $x = y = 0$, i.e. at the point $(0,0)$.

Checking for max or min:

$$\frac{\partial^2 f(0,0)}{\partial x^2} = 2 > 0;$$

$$\frac{\partial^2 f(0,0)}{\partial y^2} = 4; \quad \frac{\partial^2 f(0,0)}{\partial x \partial y} = 2;$$

$$\frac{\partial^2 f(0,0)}{\partial x^2} \cdot \frac{\partial^2 f(0,0)}{\partial y^2} - \left(\frac{\partial^2 f(0,0)}{\partial x \partial y}\right)^2 = 8 - 4 = 4 > 0;$$

According to the theorem on the prior page, since both quantities are greater than 0, the point $(0,0)$ is a minimum.

Find the critical points and determine if they are maxima or minima:

$$(x - 1)^2 + 2x(y + 2)^2$$

Take the first derivatives and set them equal to 0 to find the critical points:

$$\frac{\partial f}{\partial x} = 2x - 2 + 2(y + 2)^2 \equiv 0 \rightarrow$$
$$x = 2 - (y + 2)^2;$$

$$\frac{\partial f}{\partial y} = 4xy + 8x \equiv 0 \rightarrow y = -2;$$

Plugging $y = -2$ into the first equation gives $x = 2$. The critical point is $(2, -2)$.

Checking for max or min:

$$\frac{\partial^2 f(2,-2)}{\partial x^2} = 2 > 0;$$

$$\frac{\partial^2 f(2,-2)}{\partial y^2} = 4x = 8; \quad \frac{\partial^2 f(2,-2)}{\partial x \partial y} = 4y + 8 = 0;$$

$$\frac{\partial^2 f(2,-2)}{\partial x^2} \cdot \frac{\partial^2 f(2,-2)}{\partial y^2} - \left(\frac{\partial^2 f(2,-2)}{\partial x \partial y}\right)^2 = 16 > 0;$$

The point $(2, -2)$ is a minimum.

Chapter 5: Multiple Integrals

Single Integrals of Multi-Variable Functions:

Just as you treated extraneous independent variables as constants for partial derivatives, you do the same with integrals of multi-variable functions.

Example

Evaluate $\int x^2 y + y^2 x \, dx$.

We are taking the integral with respect to x, so we treat y as a constant:

$$\int_0^1 x^2 y + y^2 x \, dx = \frac{x^3 y}{3} + \frac{y^2 x^2}{2} \Big|_0^1 = \frac{2y+3y^2}{6}.$$

Multiple Integrals:

Similar to how you can take a derivative with respect to one variable at a time, you can take an integral with respect to one variable at a time. Sometimes it is desirable to integrate with respect to more than one independent variable. In this case, we stack the integrals and perform **partial integration** one by one. Such stacked integrals are called **iterated integrals** or **multiple integrals**.

Example

Evaluate $\int_0^1 \int_0^1 x^2 y + y^2 x \, dx dy$.

We take the integrals one at a time:

$$\int_0^1 \int_0^1 x^2 y + y^2 x \, dx dy = \int \left[\int x^2 y + y^2 x \, dx \right] dy$$

$$= \int_0^1 \frac{x^3 y}{3} + \frac{x^2 y^2}{2} \Big|_0^1 dy$$

$$= \int_0^1 \frac{2y+3y^2}{6} dy = \frac{y^2+y^3}{6} \Big|_0^1 = \frac{1}{3}.$$

Properties of Integrals:

The properties for multiple integrals follow directly from the properties you already know. They are summarized here for convenience.

Interchange Order of Integration (for definite integrals)

$$\int\left[\int f(x,y)dx\right]dy = \int\left[\int f(x,y)dy\right]dx$$

Pull Out Constants

$$\iint cf(x,y)dxdy = c\iint f(x,y)dxdy$$

Distribution

$$\iint[f(x,y) \pm g(x,y)]dxdy =$$
$$\iint f(x,y)dxdy \pm \iint g(x,y)dxdy$$

Break Intervals

$$\iint_R f(x,y)dxdy =$$
$$\iint_{R_1} f(x,y)dxdy + \iint_{R_2} f(x,y)dxdy$$

where region, R, is made up of non-overlapping regions R_1, R_2

Tools of integration, like functions of functions and integration by parts, are also just as valid as before:

Functions of Functions

$$F(u) = \int f(u)du$$

Integration by Parts

$$\int u\,dv = uv - \int v\,du$$

Example

Evaluate $\int_0^\pi \int_0^\pi y \sin 2xy \ dxdy$.

For the first partial integral with respect to x:

Let $u = 2xy; \ du = 2y$

$$\int_0^\pi \int_0^\pi y \sin 2xy \ dxdy = \frac{-1}{2} \int_0^\pi \cos 2xy \mid_0^\pi dy$$

$$= \frac{-1}{2} \int_0^\pi \cos 2\pi y - 1 \ dy$$

For the second partial integral with respect to y:

Let $u = 2\pi y; \ du = 2\pi$

$$= \frac{-1}{2} \int_0^\pi \cos 2\pi y - 1 \ dy = \frac{- \sin 2\pi y}{4\pi} + \frac{y}{2} \mid_0^\pi$$

$$= \frac{- \sin 4\pi^2}{4\pi} + \frac{\pi}{2} = 1.5.$$

Important Note: Though we treat the other variables as constants in the integration, we cannot add in these other variables to satisfy functions of functions integration. In other words, if the integral above were:

$$\int_0^\pi \int_0^\pi \sin 2xy \ dxdy$$

with $u = 2xy; \ du = 2y$

it would not be appropriate to do this:

$$\neq \int_0^\pi \left(\frac{1}{2y}\right) \int_0^\pi 2y \sin 2xy \ dxdy.$$

Other integration techniques would be needed to solve this problem.

As a final example, we consider the case of integration with functions as upper and lower limits. For these types of integrals, the order of integration is important. Work from the inside out.

Evaluate $\int_0^1 \int_y^{y^2} (1 + xy)\, dx dy$

$$\int_0^1 \int_{y^2}^y (1 + xy)\, dx dy = \int_0^1 \left[x + \frac{x^2 y}{2} \right]\Big|_{y^2}^y \, dy$$

$$= \int_0^1 \left[y + \frac{y^3}{2} - y^2 - \frac{y^5}{2} \right] dy$$

$$= \frac{y^2}{2} + \frac{y^4}{8} - \frac{y^3}{3} - \frac{y^6}{12}\Big|_0^1 = \frac{1}{2} + \frac{1}{8} - \frac{1}{3} - \frac{1}{12} = \frac{5}{24}$$

If we wanted to reverse the order of integration, we would have to rework all of the upper and lower limits. In this example, the limits are: $0 \leq y \leq 1$ and the functions $x = y^2$ and $x = y$.

As per the graph to the right, the new limits are:
$0 \leq x \leq 1$; $y = x$ and $y = \sqrt{x}$.

The new integral becomes:

$$\int_0^1 \int_x^{\sqrt{x}} (1 + xy)\, dy dx = \int_0^1 \left[y + \frac{y^2 x}{2} \right]\Big|_x^{\sqrt{x}} \, dx$$

$$= \int_0^1 \left[\sqrt{x} + \frac{x^2}{2} - x - \frac{x^3}{2} \right] dx$$

$$= \frac{2\sqrt{x}^3}{3} + \frac{x^3}{6} - \frac{x^2}{2} - \frac{x^4}{8}\Big|_0^1 = \frac{5}{24}$$

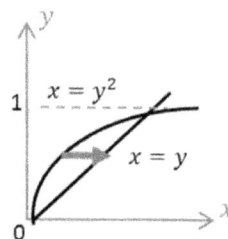

Limits of integration for the example. The arrow shows lower to upper bounds for the function limits.

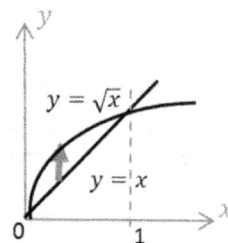

Reversing the order of integration requires careful attention to the limits. Note the change in the arrow and reversal of upper and lower limits.

Sometimes it is necessary to split the area into regions just as was done in Calculus 1 for single variable integration. Problem 5.14 gives an example of region splitting.

Chapter 6: Cylindrical Coordinates

Cartesian Coordinates:

Before tackling new coordinate systems, let's first review the familiar system we've been using to this point: the **Cartesian (or Rectangular) coordinate system**.

The Cartesian coordinate system describes all points (x, y, z) as though they exist in a rectangular block, with x being one side, y another side, and z the final side. In other words, if you want to find a point (a, b, c), you would travel a distance a in the x direction, b in the y direction, and c in the z direction.

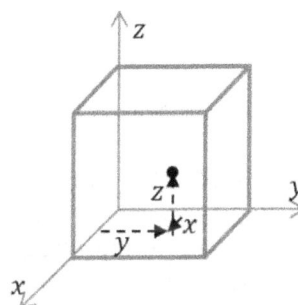

Right Handed Cartesian Coordinate System.

It should also be noted that for our mathematics to work properly, we use a **right handed Cartesian coordinate system**. This means that if you point the thumb of your right hand in the $+z$ direction, when your fingers are straight, they point in the $+x$ direction, and when you curl your fingers to point 90° away, they point in the $+ y$ direction.

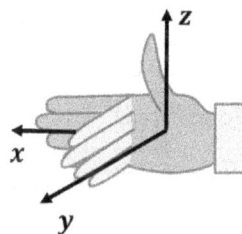

Employing the Right Hand Rule.

Calculus in Cartesian Coordinates:

When you perform an integration or find a derivative, you deal in quantities called dx, dy, and dz. These differential elements are small units of distance along their respective axes, thus they have units of length. As you've seen in Calculus 1, it is important that these differential elements have units of length in order to use the integral to calculate area. The **differential elements of integration** in Cartesian coordinates are:

$$dx, \quad dy, \quad dz.$$

And, the **derivative operators** are:

$$\frac{\partial}{\partial x}, \quad \frac{\partial}{\partial y}, \quad \frac{\partial}{\partial z}.$$

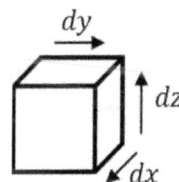

Differential elements for Cartesian coordinates.

Cylindrical Coordinates:

There are other ways to find the point (x, y, z) by using other coordinate systems. The **Cylindrical** (or **Polar** in the $x - y$ plane) **coordinate system** describes points based on their location on a cylinder.

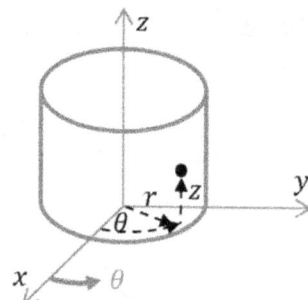

Cylindrical Coordinate System.

The point (x, y, z) in Rectangular coordinates can also be expressed in terms of (r, θ, z) where we have a radius, r, an angle, θ, and a height, z. To find this point, you go out a distance r in the $x - y$ plane at an angle of θ to the x-axis, then go up a distance z.

To change from Cartesian to Cylindrical coordinates, we use trigonometry:

$$r = \sqrt{x^2 + y^2}; \quad \theta = \tan^{-1}\left(\frac{y}{x}\right); \quad z = z.$$

To go from Cylindrical to Cartesian coordinates, we use:

$$x = r\cos\theta; \quad y = r\sin\theta; \quad z = z.$$

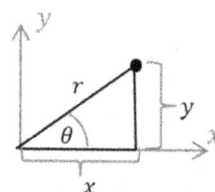

Geometry for converting between (x, y) and (r, θ).

Example

Express $(x, y, z) = (3, 4, 8)$ in Cylindrical coordinates:

$r = \sqrt{3^2 + 4^2} = \sqrt{25} = 5;$
$\theta = \tan^{-1}\frac{4}{3} = 0.93; \ z = 8;$

The point becomes $(5, 0.93, 8)$.

Example

Express $(r, \theta, z) = (4, \frac{\pi}{2}, 10)$ in Rectangular coordinates:

$x = 4\cos\frac{\pi}{2} = 0; \ y = 4\sin\frac{\pi}{2} = 4; \ z = 10;$

The point becomes $(0, 4, 10)$.

Calculus in Cylindrical Coordinates:

As mentioned previously, integration is meant to be performed against units of length. Whereas, x, y, z, and r all have units of length, θ does not.

We can illustrate this problem by considering the differential element $\Delta\theta$. Let's assume $\Delta\theta$ is 1° or 17.5 mrad (milliradians). Using trigonometry to find the small distance Δl as shown to the right, we would say that:

$$\Delta l = r \sin \Delta\theta.$$

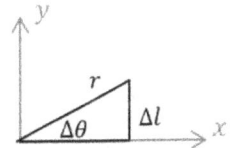

The length Δl depends on r and $\Delta\theta$.

If $r = 1$", then $\Delta l = 1.75 \times 10^{-5}$ inches. But, if $r = 100$ yards, then $\Delta l = 63$ inches. So, we have to know what r is in order to know what our differential length is. Using the above expression as $\Delta\theta \to d\theta$, we get:

$$dl = r \sin d\theta \approx r d\theta.$$

In the last step, we used the **paraxial approximation** that the sine of a very small angle is approximately the value of that angle.

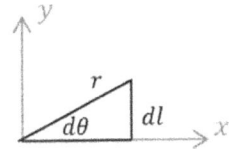

$dl = r \sin d\theta \sim r d\theta$

Finding a differential element for $d\theta$.

The bottom line is when you integrate with respect to r or z, you can use dr or dz. But, when you integrate with respect to θ, you must use $r d\theta$. The **differential elements of integration** in Cylindrical coordinates are:

$$dr, \quad r d\theta, \quad dz.$$

To take derivatives, one reverses the process of integration. Derivatives with respect to r and z are performed just like those with x and y. However, the θ derivative must be divided by r after the differentiation. The **derivative operators** are:

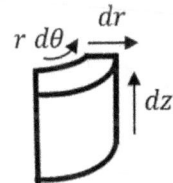

Differential elements for Polar coordinates.

$$\frac{\partial f}{\partial r}, \quad \frac{1}{r}\frac{\partial f}{\partial \theta}, \quad \frac{\partial f}{\partial z}.$$

Integrate the function over the region specified:

Example

$$f(r,\theta) = r^2; \ 0 \le r \le 2; \ 0 \le \theta \le \pi;$$

The geometry of the problem is shown to the right.

$$\int_0^\pi \int_0^2 f(r,\theta)\, r\, dr\, d\theta = \int_0^\pi \int_0^2 r^3 \, dr\, d\theta$$

$$= \int_0^\pi \frac{r^4}{4}\Big|_0^2 \, d\theta = \int_0^\pi 4 \, d\theta = 4\theta\big|_0^\pi = 4\pi.$$

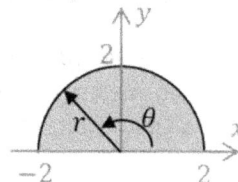

Region of integration for the example problem.

Reversing the order of integration yields the same result (see side note).

If we leave the answer in symbolic form, $\frac{r^4\theta}{4}$, we can take the derivative to obtain the original function:

$$\frac{1}{r}\frac{\partial}{\partial\theta}\left[\frac{\partial}{\partial r}\left(\frac{r^4\theta}{4}\right)\right] = \frac{1}{r}\frac{\partial}{\partial\theta}[r^3\theta] = \frac{r^3}{r} = r^2$$

Reversing the order of integration:

$$\int_0^2 \int_0^\pi f(r,\theta)\, r\, d\theta\, dr =$$
$$\int_0^2 \int_0^\pi r^3 \, d\theta\, dr =$$
$$\int_0^2 \theta r^3 \big|_0^\pi \, dr = \int_0^2 \pi r^3 \, dr =$$
$$\frac{\pi r^4}{4}\Big|_0^2 = 4\pi.$$

Integrate the function over the region specified:

Example

$$f(r,\theta) = r\cos\theta; \ 0 \le r \le 1; \ 0 \le \theta \le \pi;$$

If we don't split regions, we get an answer of 0. We can predict this because both the region and the θ function $(\cos\theta)$ are symmetric about the y axis.

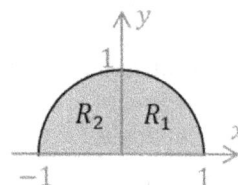

Regions of integration for the example problem.

$$\int_0^\pi \int_0^1 f(r,\theta)\, r\, dr\, d\theta$$

$$= \int_0^{\pi/2}\int_0^1 r^2 \cos\theta \, dr\, d\theta + \int_{\pi/2}^\pi \int_0^1 r^2 \cos\theta \, dr\, d\theta$$

$$= \int_0^{\pi/2} \frac{r^3}{3}\cos\theta\, \big|_0^1 \, d\theta += \int_{\pi/2}^\pi \frac{r^3}{3}\cos\theta\, \big|_0^1 \, d\theta$$

$$= \int_0^{\pi/2} \frac{\cos\theta}{3} \, d\theta + \int_{\pi/2}^\pi \frac{\cos\theta}{3} \, d\theta$$

$$= \frac{\sin\theta}{3}\Big|_0^{\pi/2} + \frac{\sin\theta}{3}\Big|_{\pi/2}^\pi = \frac{1}{3} - \frac{-1}{3} = \frac{2}{3}.$$

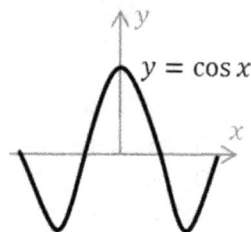

Cosine is also symmetric about the y axis, so split the region of integration.

Chapter 7: Spherical Coordinates

Spherical Coordinates:

The **Spherical coordinate system** describes points based on their location on a sphere.

The point (x, y, z) can be expressed in terms of (ρ, φ, θ) where we have a radius, ρ, and two angles, θ and φ. To find this point, you go out a distance ρ from the origin at an angle of θ to the x-axis and φ to the z-axis.

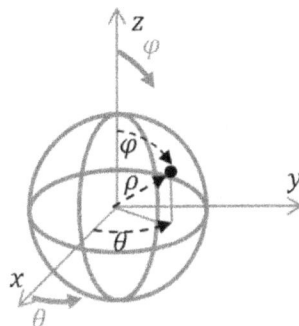

Spherical Coordinate System.

To change from Cartesian to Spherical coordinates, we again use trigonometry. We see from the figure that r is the **projection** (or shadow) of ρ onto the $x - y$ plane where the relations

$$r = \sqrt{x^2 + y^2} \text{ and } \theta = \tan^{-1}\left(\frac{y}{x}\right)$$

still hold. To determine ρ and φ, the triangle comprised of the lengths r, ρ, and z gives us the following relationships:

$$\rho = \sqrt{r^2 + z^2}; \ \varphi = \tan^{-1}\left(\frac{r}{z}\right).$$

Plugging in our expression for r, we get:

$$\rho = \sqrt{x^2 + y^2 + z^2}; \varphi = \tan^{-1}\left(\frac{\sqrt{x^2+y^2}}{z}\right); \theta = \tan^{-1}\left(\frac{y}{x}\right).$$

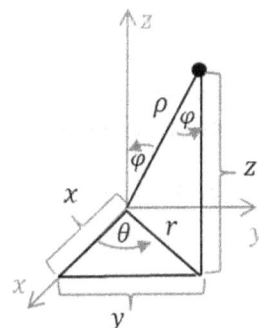

$$r = \sqrt{x^2 + y^2} = \rho \sin\varphi$$
$$\rho = \sqrt{r^2 + z^2}$$
$$\varphi = tan^{-1}\left(\frac{r}{z}\right)$$

Geometry for converting between (x, y, z) and (r, φ, θ).

To go from Spherical to Cartesian coordinates, we use:

$$x = \rho \sin\varphi \cos\theta; \ \ y = \rho \sin\varphi \sin\theta; \ \ z = \rho \cos\varphi.$$

Example

Express $(x, y, z) = (3, 4, 8)$ in Spherical coordinates:

$$\rho = \sqrt{3^2 + 4^2 + 8^2} = 9.4;$$

$$\theta = \tan^{-1}\frac{4}{3} = 0.9; \ \varphi = \tan^{-1}\left(\frac{\sqrt{3^2+4^2}}{8}\right) = 0.5;$$

The point becomes $(9.4, 0.9, 0.5)$.

Calculus in Spherical Coordinates:

Integration is again modified slightly to convert angles to units of length. We first consider the angle φ off of the z-axis. Performing the same action we did in Polar coordinates:

$$dl_1 = \rho \sin d\varphi \approx \rho d\varphi.$$

Now we consider the angle, θ, from the x-axis. We need the arm length, r, to determine the differential length, dl_2 as defined by $d\theta$. The geometry of the problem is given to the right.

For a given ρ and φ, r is the projection of ρ onto the $x - y$ plane. By trigonometry, r has a length $\rho \sin \varphi$. (Since the magnitude of φ can be anything for a given function, we can't use the paraxial approximation for $\sin \varphi$.)

With $d\theta$ being the angle from the x-axis to r, the differential length, dl_2, becomes:

$$dl_2 = r \sin d\theta = [\rho \sin \varphi] \sin d\theta \approx \rho \sin \varphi \, d\theta.$$

Here, we *can* use the paraxial approximation for $\sin d\theta$ since we define $d\theta$ to be small.

So, in summary, the **differential elements of integration** in Spherical coordinates are:

$$d\rho, \quad \rho d\varphi, \quad \rho \sin \varphi \, d\theta.$$

Reversing the process gives the **derivative operators**:

$$\frac{\partial}{\partial \rho}, \quad \frac{1}{\rho}\frac{\partial}{\partial \varphi}, \quad \frac{1}{\rho \sin \varphi}\frac{\partial}{\partial \theta}.$$

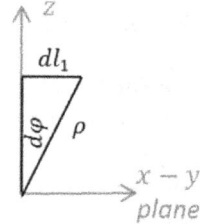

$dl_1 = \rho \sin d\varphi \sim \rho d\varphi$

The length dl_1 depends on ρ and $d\varphi$.

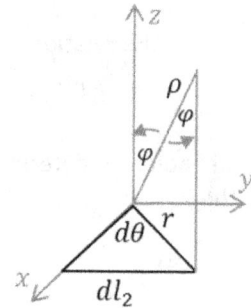

$dl_2 = r \sin d\theta \sim \rho \sin \varphi \, d\theta$

Finding a differential element for $d\theta$.

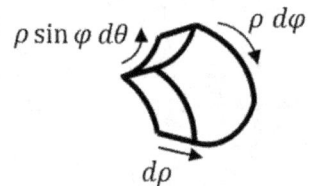

Differential elements for Spherical coordinates.

Integrate the function over the region specified:

$$f(\rho, \varphi, \theta) = \rho^2; \ \ 0 \le \rho \le 2; \ \ 0 \le \varphi \le \pi;$$

Here, the variables of integration are и ρ, φ:
$$\int_0^\pi \int_0^2 f(\rho, \varphi, \theta) \, \rho d\rho d\varphi = \int_0^\pi \int_0^2 \rho^3 \, d\rho d\varphi$$

Reversing the order of integration:

$$\int_0^2 \int_0^\pi f(\rho, \varphi, \theta) \, \rho d\varphi d\rho =$$

$$= \int_0^\pi \frac{\rho^4}{4} \Big|_0^2 d\varphi = \int_0^\pi 4 \, d\varphi = 4\varphi \Big|_0^\pi = 4\pi.$$

$$\int_0^2 \int_0^\pi \rho^3 \, d\varphi d\rho =$$

Reversing the order of integration yields the same result (see side note). If we leave the answer in symbolic form, $\frac{\rho^4 \varphi}{4}$, the derivative will recover the original function:

$$\int_0^2 \varphi\rho^3 \Big|_0^\pi d\rho = \int_0^2 \pi\rho^3 d\rho$$

$$\frac{\pi\rho^4}{4} \Big|_0^2 = 4\pi.$$

$$\frac{1}{\rho} \frac{\partial}{\partial \varphi} \left[\frac{\partial}{\partial \rho} \left(\frac{\rho^4 \varphi}{4} \right) \right] = \frac{1}{\rho} \frac{\partial}{\partial \varphi} [\rho^3 \varphi] = \frac{\rho^3}{\rho} = \rho^2.$$

Integrate the function over the region specified:

$$f(\rho, \varphi, \theta) = \rho^2; \ 0 \le \rho \le 2; \ 0 \le \theta \le \pi;$$

Here, the variables of integration are и ρ, θ:
$$\int_0^\pi \int_0^2 f(\rho, \varphi, \theta) \, \rho \sin \varphi \, d\rho d\theta$$

Reversing the order of integration:

$$\int_0^2 \int_0^\pi f(\rho, \varphi, \theta) \, \rho \sin\varphi d\theta d\rho$$

$$= \int_0^\pi \int_0^2 \rho^3 \sin \varphi \, d\rho d\theta = \int_0^\pi \frac{\rho^4}{4} \sin \varphi \Big|_0^2 d\theta$$

$$= \int_0^2 \int_0^\pi \rho^3 \sin \varphi \, d\theta d\rho =$$

$$= \int_0^\pi 4 \sin \varphi \, d\theta = 4\theta \sin \varphi \Big|_0^\pi = 4\pi \sin \varphi.$$

$$\int_0^2 \theta\rho^3 \sin \varphi \Big|_0^\pi d\rho$$

If we leave the answer in symbolic form, $\frac{\rho^4 \theta \sin \varphi}{4}$, the derivative will recover the original function:

$$= \int_0^2 \pi\rho^3 \sin \varphi \, d\rho =$$

$$\frac{1}{\rho \sin \varphi} \frac{\partial}{\partial \theta} \left[\frac{\partial}{\partial \rho} \left(\frac{\rho^4 \theta \sin \varphi}{4} \right) \right]$$

$$\frac{\pi\rho^4 \sin \varphi}{4} \Big|_0^2 = 4\pi \sin \varphi.$$

$$= \frac{1}{\rho \sin \varphi} \frac{\partial}{\partial \theta} [\rho^3 \theta \sin \varphi] = \frac{\rho^3 \sin \varphi}{\rho \sin \varphi} = \rho^2.$$

Coordinate System Summary:

Cartesian

		Convert to Cylindrical	**Convert to Spherical**
$\dfrac{\partial}{\partial x}$	dx	$r = \sqrt{x^2 + y^2}$	$\rho = \sqrt{x^2 + y^2 + z^2}$
$\dfrac{\partial}{\partial y}$	dy	$\theta = \tan^{-1}\left(\dfrac{y}{x}\right)$	$\varphi = \tan^{-1}\left(\dfrac{\sqrt{x^2+y^2}}{z}\right)$
$\dfrac{\partial}{\partial z}$	dz	$z = z$	$\theta = \tan^{-1}\left(\dfrac{y}{x}\right)$

Cylindrical

		Convert to Cartesian	**Convert to Spherical**
$\dfrac{\partial}{\partial r}$	dr	$x = r\cos\theta$	$\rho = \sqrt{r^2 + z^2}$
$\dfrac{1}{r}\dfrac{\partial}{\partial \theta}$	$rd\theta$	$y = r\sin\theta$	$\varphi = \tan^{-1}\left(\dfrac{r}{z}\right)$
$\dfrac{\partial}{\partial z}$	dz	$z = z$	$\theta = \theta$

Spherical

		Convert to Cylindrical	**Convert to Cartesian**
$\dfrac{\partial}{\partial \rho}$	$d\rho$	$r = \rho\sin\varphi$	$x = \rho\sin\varphi\cos\theta$
$\dfrac{1}{\rho}\dfrac{\partial}{\partial \varphi}$	$\rho d\varphi$	$\theta = \theta$	$y = \rho\sin\varphi\sin\theta$
$\dfrac{1}{\rho\sin\varphi}\dfrac{\partial}{\partial \theta}$	$\rho\sin\varphi\, d\theta$	$z = \rho\cos\varphi$	$z = \rho\cos\varphi$

Chapter 8: Length, Area, & Volume

Area & Volume of Functions:

Recall that in single-variable calculus, the integral $\int f(x)\,dx$ gives the **area** under a function. In essence, the integral sums up all of the small rectangular areas with $y = f(x)$ as the height and dx as the width.

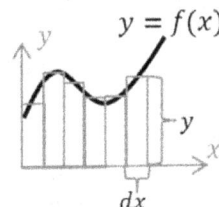

Finding area in single variable calculus.

In multi-variable calculus, the single integral gives the **cross-sectional area** of the function:

$$A_{cross-section} = \int f(x,y)dx.$$

In the above equation, we are taking a cross-section parallel to the $x - z$ plane (summing rectangles made up of $z = f(x,y)$ and dx). We end up with an expression in terms of y ($A_{cross-section} = f(y)$) which allows us to find the cross-sectional area along some line $y = c$.

Similar to the area equation, the double integral gives the **volume** of the function:

$$V = \iint f(x,y)dxdy.$$

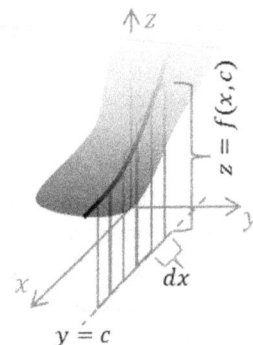

Finding area in two variable calculus.

Here, small cubes with sides $z = f(x,y)$, dx, and dy are summed in two integrals. These relations to area and volume are why it is important to have units of length in the differential elements of the integrand.

Example Find the cross-sectional area of the function $f(x,y) = yx^3$ from $0 \le x \le 2$ at $y = 1$.

$$A_{cs} = \int_0^2 yx^3 dx = \left.\frac{yx^4}{4}\right|_0^2 = 4y\,\Big|_{y=1} = 4.$$

Example Find the volume of the function $f(x,y) = yx^3$ from $0 \le x \le 2$ and $0 \le y \le 2$.

Note: The last problems of Chapters 6 & 7 gave examples of this type. The last problems for this chapter give practice with triple integrals.

$$V = \int_0^2 \int_0^2 yx^3 dxdy = \int_0^2 4y\,dy = 2y^2\big|_0^2 = 8.$$

Expressions of Length, Area, & Volume:

It may surprise you that there are reasons to have integrals without functions in the integrand. It may sound like a sentence without a subject, but these expressions can be useful. In short, an integral without a function yields a result that is one dimension less than its counterpart with a function.

Let's define dL as a **differential length**, dA as a **differential area**, and dV as a **differential volume**. Then the expressions for **length**, L, **area**, A, and **volume**, V, are:

$$L = \int dL; \quad A = \iint dA; \quad V = \iiint dV.$$

Let's further define that dL can have one single-dimensional, differential element, dA can have two, and dV can have three. The choice of single-dimensional, differential elements depends on the coordinate system, but within a system, you can choose any one, two, or three as needed:

Cartesian	Cylindrical	Spherical
dx, dy, dz	$dr, rd\theta, dz$	$d\rho, \rho d\varphi, \rho \sin\varphi \, d\theta$

Example

Find the area of a circle.

The cylindrical coordinate system lends itself well to describing a circle with r and θ. Our differential elements are then dr and $rd\theta$. We draw the circle by going a distance R from the origin and looping one full turn, or 2π. The limits of integration are:

$$0 \leq r \leq R, 0 \leq \theta \leq 2\pi$$

$$\int_0^{2\pi} \int_0^R r \, dr \, d\theta = \int_0^{2\pi} \frac{R^2}{2} d\theta = \frac{R^2 \theta}{2} \Big|_0^{2\pi} = \pi R^2.$$

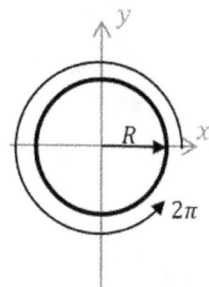

Finding the area of a circle.

Example

Find the cross-sectional area of a sphere.

The spherical coordinate system seems a good choice. We can choose ρ and φ or ρ and θ. Our differential elements are then $d\rho$ and $\rho d\varphi$ or $d\rho$ and $\rho \sin \varphi\, d\theta$. In either case, the cross-section is defined with:

$$0 \le r \le R, 0 \le \varphi \text{ or } \theta \le 2\pi$$

$$\int_0^{2\pi} \int_0^R \rho d\rho d\varphi = \int_0^{2\pi} \frac{R^2}{2} d\varphi = \frac{R^2 \varphi}{2}\Big|_0^{2\pi} = \pi R^2.$$

or

$$\int_0^{2\pi} \int_0^R \rho \sin \varphi\, d\rho d\theta = \int_0^{2\pi} \frac{R^2 \sin \varphi}{2} d\theta$$

$$= \frac{\theta R^2 \sin \varphi}{2}\Big|_0^{2\pi} = \pi R^2 \sin \varphi.$$

The latter gives a more complete description of the cross-sectional area of a sphere along an axis. Consider the top ($\varphi = 0$) and bottom ($\varphi = \pi$) where $A = 0$. At the midpoint ($\varphi = \pi/2$), $A = \pi R^2$. Similarly, we can find any cross-section in between.

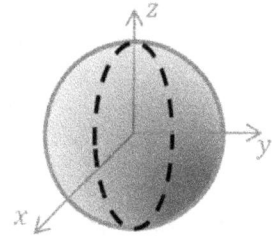

Finding the cross-sectional area of a sphere using φ.

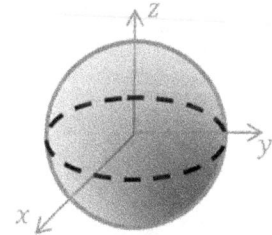

Finding the cross-sectional area of a sphere using θ.

An important note about working in spherical coordinates: Because there are two angles, people sometimes integrate both from their two extremes 0 to 2π. If you do this, you actually double count. If you are evaluating a function in the full angular φ and θ, you should evaluate them from 0 to π and 0 to 2π, respectively.

Example

Find the volume of a sphere.

$$\int_0^{2\pi} \int_0^\pi \int_0^R \rho^2 \sin \varphi\, d\rho d\varphi d\theta$$

$$= \int_0^{2\pi} \int_0^\pi \frac{R^3 \sin \varphi}{3} d\varphi d\theta = \int_0^{2\pi} \frac{-R^3 \cos \varphi}{3}\Big|_0^\pi d\theta$$

$$= \int_0^{2\pi} \frac{2R^3}{3} d\theta = \frac{2\theta R^3}{3}\Big|_0^{2\pi} = \frac{4\pi R^3}{3}.$$

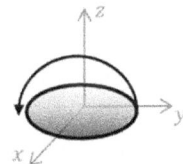

Once a circle is drawn in the $x - y$ plane ($\theta = 2\pi$), it only needs to flip once ($\varphi = \pi$) to outline a sphere.

Part 2: Limits, Sequences, & Series

Chapter 9: Limits with Indeterminate Forms

Properties of Limits:

In general, limits have some nice properties that make them rather straightforward to evaluate. The assumption of this text is that basic theories on limits have been covered in prior mathematics courses. A summary of limit properties is given below for convenience:

Pull Out Constants

$$\lim_{x \to a} cf(x) = c\left[\lim_{x \to a} f(x)\right]$$

Distribution

$$\lim_{x \to a} f(x) \pm g(x) = \lim_{x \to a} f(x) \pm \lim_{x \to a} g(x)$$

Product Rule

$$\lim_{x \to a} f(x) \cdot g(x) = \left[\lim_{x \to a} f(x)\right] \cdot \left[\lim_{x \to a} g(x)\right]$$

Quotient Rule

$$\lim_{x \to a} \frac{f(x)}{g(x)} = \frac{\lim_{x \to a} f(x)}{\lim_{x \to a} g(x)}$$

Power Rule

$$\lim_{x \to a} f(x)^n = \left[\lim_{x \to a} f(x)\right]^n$$

---

*   *For the remainder of this book, we return again to single-variable functions. The concepts can be applied directly to functions with multiple variables.*

Limits with Indeterminate Forms:

There are times when limits yield answers that are questionable because they involve some relation with infinity, a divide by zero, or similar. These limits are said to have **indeterminate forms**. Though the initial calculation may give such results, further techniques can be employed for a better approximation. Three special cases are outlined below.

$L = 0/0$ or ∞/∞:

If you perform a limit on $f(x) = \frac{g(x)}{h(x)}$ that gives $\frac{0}{0}$ or $\frac{\infty}{\infty}$, then use **L'Hôptial's Rule**:

$$\text{If } L = \lim_{x \to a} f(x) = \lim_{x \to a} \frac{g(x)}{h(x)} = \frac{0}{0} \, or \, \frac{\infty}{\infty},$$

$$\text{Then } L = \lim_{x \to a} \frac{g'(x)}{h'(x)}.$$

There are some obvious caveats to the rule like $g'(x)$ and $h'(x)$ exist and $h'(x) \neq 0$. The possible exception to that is at $x = a$. Also, you need to get a final answer that exists or is equal to infinity for it to be valid.

Example Find $\lim_{x \to -1} \frac{4x^2 - 4}{x+1}$;   The limit has form $\frac{0}{0}$

$$\lim_{x \to -1} \frac{4x^2-4}{x+1} = \lim_{x \to -1} \frac{8x}{1} = 8.$$

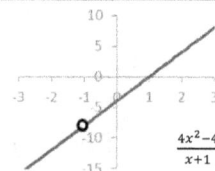

$\frac{4x^2-4}{x+1}$

Example Find $\lim_{x \to \infty} \frac{e^{2x}}{3x^2}$;   Form is $\frac{\infty}{\infty}$

$$= \lim_{x \to \infty} \frac{2e^{2x}}{6x} \quad \text{Form is still } \frac{\infty}{\infty}, \text{ so we do it again:}$$

$$= \lim_{x \to \infty} \frac{4e^{2x}}{6} = \frac{\infty}{6} = \infty.$$

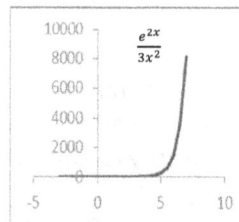

$\frac{e^{2x}}{3x^2}$

$L = 0 \cdot \infty$:

If you perform a limit where you get $L = 0 \cdot \infty$, you can use a simple division trick so that you can then use L'Hôptial's Rule:

> **If $L = \lim_{x \to a} f(x) = \lim_{x \to a} g(x)h(x) = 0 \cdot \infty$,**
>
> **Then $L = \lim_{x \to a} \dfrac{D_x[g(x)]}{D_x\left[\frac{1}{h(x)}\right]}$ or $L = \lim_{x \to a} \dfrac{D_x[h(x)]}{D_x\left[\frac{1}{g(x)}\right]}$.**

You can use $1/h(x)$ or $1/g(x)$, whatever is most convenient. The same rules apply as before with L'Hôptial's Rule for $g(x)$ and $1/h(x)$ or $1/g(x)$ and $h(x)$.

Example

Find $\lim_{x \to 0} x^3 \ln x$; Form is $\infty \cdot 0$

We use the relation $x^3 = \dfrac{1}{x^{-3}}$:

$$\lim_{x \to 0} x^3 \ln x = \lim_{x \to 0} \frac{\ln x}{x^{-3}}$$

Now, the limit has the form $\dfrac{0}{0}$, and we can use L'Hôptial's Rule as usual:

$$\lim_{x \to 0} \frac{x^{-1}}{-3x^{-4}} = \lim_{x \to 0} -\frac{1}{3}x^3 = 0$$

Always be careful to check the form before employing any of these techniques, or you may get tricked into a wrong answer.

$L = 0^0, \infty^0, \text{or } 1^\infty$:

Though at first blush it would appear that expressions like $0^0, \infty^0$, or 1^∞ have obvious values, if you get such a result from a limit of $f(x) = g(x)^{h(x)}$, then use the properties of logarithms to find a better approximation:

If $L = \lim_{x \to a} f(x) = \lim_{x \to a} g(x)^{h(x)} = 0^0, \infty^0, \text{or } 1^\infty,$

Then $L_{new} = \lim_{x \to a}[\ln f(x)] = \lim_{x \to a}[h(x) \ln g(x)]$

And, $L = \lim_{x \to a} f(x) = e^{L_{new}}.$

In other words:
   If $\ln f(x)$ has limit L_{new},
   Then $f(x)$ has limit $L = e^{L_{new}}$.

Example

Find $\lim_{x \to -1} (4x^2 - 4)^{(x+1)};$ The limit has form 0^0

$L_{new} = \lim_{x \to -1} \ln(4x^2 - 4)^{(x+1)}$

$= \lim_{x \to -1} (x + 1) \ln(4x^2 - 4)$ Form is $0 \cdot \infty$

$= \lim_{x \to -1} \dfrac{\ln(4x^2 - 4)}{(x+1)^{-1}}$ Form is $\dfrac{\infty}{\infty}$

$= \lim_{x \to -1} \dfrac{8x/(4x^2 - 4)}{-(x+1)^{-2}} = \lim_{x \to -1} \dfrac{-8x(x+1)^2}{4(x+1)(x-1)}$

$= \lim_{x \to -1} \dfrac{-2x(x+1)}{(x-1)} = \dfrac{2 \cdot 0}{-2} = 0$

Don't forget the last step!

$\lim_{x \to -1} (4x^2 - 4)^{(x+1)} = L = e^{L_{new}} = e^0 = 1.$

Chapter 10: Infinite Sequences

Terminology:

Infinite sequences (or just **sequences**) are simply lists of numbers or functions that follow some pattern forever. An example would be the sequence 1, 3, 5, 7, 9, 11, …. When talking about infinite sequences, there are some notations and terms you should be familiar with:

<u>Notation:</u> A single term in an infinite sequence is represented by a_n, where n denotes the nth element. A generic sequence is written as:

$$\{a_n\} = a_1, a_2, a_3, \cdots, a_n, \cdots$$

Many times, the term a_n is replaced by a formula that shows how the sequence is constructed. As an example:

$$1, 3, 5, \cdots, 2n - 1, \cdots$$

<u>Ordered:</u> Sequences are **ordered** meaning their terms occur in a certain order.

<u>Recursive Definition:</u> If you are given the first term, a_1, and the formula for finding the next term based on the prior term (i.e., $a_k = f(a_{k-1})$), then you can figure out the nth term by inductive reasoning. Defining a sequence with the terms a_1 and a_k is called giving a **recursive definition**.

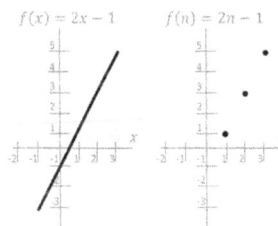

$f(x) = 2x - 1$ $f(n) = 2n - 1$

You can think of sequences as functions, $f(n)$, but instead of being valid for all values of x, they are only valid for positive integers, $n \geq 1$

> **Example**
>
> Find an expression for the nth term of the recursively defined sequence given by:
> $$a_1 = 5; \quad a_k = a_{k-1} - 2;$$
>
> Setting up a table can help to see the patterns:
>
> Remember:
> a_k is defined by the previous term a_{k-1},
> a_n is defined by the element number, n.

a_k	5	5-2= 3	3-2= 1	1-2= -1	$\cdots$	$a_{k-1} - 2$
n	1	2	3	4	$\cdots$	n
a_n	7-2(1)= 5	7-2(2)= 3	7-2(3)= 1	7-2(4)= -1	$\cdots$	$7 - 2n$

Arithmetic Sequence: The successive terms of an **arithmetic sequence** are found by adding (or subtracting) a constant. An example is again:

$$1, 3, 5, \cdots, 2n - 1, \cdots$$

where the **arithmetic constant** here is 2 (the sequence is defined recursively as $a_1 = 1$; $a_k = 2 + a_{k-1}$).

Geometric Sequence: The successive terms of a **geometric sequence** are found by multiplying (or dividing) by a constant. An example is:

$$2, 4, 8, 16, \cdots, 2^n, \cdots$$

where the **geometric constant** here is 2 (the sequence is defined recursively as $a_1 = 2$; $a_k = 2 \times a_{k-1}$).

Alternating Sequence: The successive terms of an **alternating sequence** change sign. An example is the alternating, geometric sequence with constant -2:

$$-2, +4, -8, +16, \cdots, (-1)^n(2^n), \cdots$$

Monotonic Sequence: A sequence is **monotonic** if every successive term always increases or always decreases. Our repeated example of the odd integers is an increasing, monotonic, arithmetic sequence.

Bounded Sequence: If a sequence is **bounded**, that means that as $n \to \infty$, no terms in the sequence exceed an upper and/or lower limit.

Convergent Sequence: If a sequence is **convergent**, then as $n \to \infty$, the terms in the sequence get closer and closer to a particular value. The burning question with sequences is whether they are convergent, or the opposite, **divergent**, which is a nice segue to the next section – after an example...

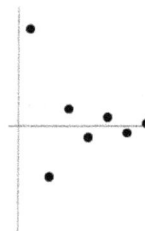

Alternating Sequence Converging to 0.

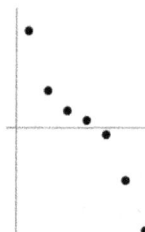

Decreasing Monotonic Sequence

$f(n) = 1 + (n - 4)^2$

Lower Bounded Sequence

For the following sequence, find the recursive definition , determine the arithmetic or geometric constant if there is one, and state the sequence properties:

$\{3^n + 3^{n-1}\}$

The first terms are:

n	1	2	3	4
a_n	$3^1 + 3^0$ 4	$3^2 + 3^1$ 12	$3^3 + 3^2$ 36	$3^4 + 3^3$ 108

Looking at these terms, you may notice that starting with $a_1 = 4$, each successive term is the prior one multiplied by 3: $a_k = 3 \times a_{k-1}$.

This is a geometric sequence with a geometric constant of 3 (based on the recursive definition given above).

The sequence appears to always be increasing from the initial value of 4 to infinity. Therefore, we have:
- a monotonically increasing,
- lower bounded (lower bound is 4),
- divergent (goes to infinity),
- geometric sequence.

Sequence Convergence

There is a formal definition of sequence convergence:

> **A sequence converges to $L = \lim_{n \to \infty} a_n$**
> **if $|a_n - L| < \varepsilon$ for $n > N$ and $\varepsilon > 0$.**

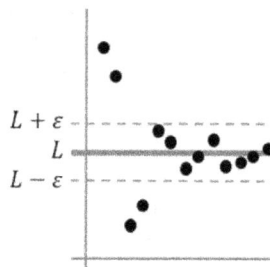

This is best explained by the picture to right. Basically, it means that the later terms of a convergent series are bounded by $L \pm \varepsilon$ where ε is some positive number.

No matter what happens with the fist terms, the later terms of a convergent sequence all hover around the value L with some error, $\pm \varepsilon$

There are a few tests for convergence. The first is:

> **Let $f(n) = a_n$ and let $f(x)$ exist for $x \geq 1$, then**
> **If $\lim_{x \to \infty} f(x) = L$, then $\lim_{n \to \infty} f(n) = L$ where $L \in (-\infty, \infty)$.**

All this means is plug in x for n and find the limit.

The second is called the **sandwich theorem**:

> **If $a_n \leq b_n \leq c_n \ \forall \, n$, and $\lim_{n \to \infty} a_n = \lim_{n \to \infty} c_n = L$,**
> **then $\lim_{n \to \infty} b_n = L$.**

If a sequence is always bounded by two other sequences that converge to the same value, then the bounded sequence converges to that value, as well.

The third states simply:

> **A bounded, monotonic, infinite sequence converges.**

Note that it must be upper bounded for increasing sequences and lower bounded for decreasing ones.

Finally, for geometric sequences (note "r" is a constant):

> $$\lim_{n \to \infty} r^n = 0, |r| < 1; \quad \lim_{n \to \infty} |r^n| = \infty, |r| > 1.$$

> **In Summary**
>
> Ways to determine sequence convergence:
>
> - Take the limit of a_n
>
> - Check if it is "sandwiched"
>
> - Check if it's bounded & monotonic
>
> - Check if it's a geometric sequence

Example

Determine if the sequence converges: $\left\{\frac{4n^2+2}{n^3}\right\}$

Using the first test of convergence, we simply take the limit of the nth term:

$$\lim_{x\to\infty}\frac{4x^2+2}{x^3}=\frac{\infty}{\infty};\text{ We can use L'Hôptial's Rule:}$$

$$\lim_{x\to\infty}\frac{4x^2+2}{x^3}=\lim_{x\to\infty}\frac{8x}{3x^2}=\lim_{x\to\infty}\frac{8}{6x}=0$$

The sequence converges to 0.

Example

Determine if the sequence converges: $\{e^{-n}\}$

The sequence can also be written as $\left\{\left(\frac{1}{e}\right)^n\right\}$. This is a geometric sequence of the form $\{r^n\}$ with $|r|<1$, therefore the sequence converges to 0.

Example

Determine if the sequence converges: $\left\{\frac{\sin^{2n}n}{8^n}\right\}$

Since we know that $-1\le \sin(x)\le 1$, we also know that even powers of $\sin(x)$ are bounded by 0 and 1.

$$0\le \sin^{2n}(n)\le 1$$

$$0\le \frac{\sin^{2n}n}{8^n}\le \frac{1}{8^n}$$

$$\lim_{n\to\infty}0=0;\qquad \lim_{n\to\infty}\left(\frac{1}{8}\right)^n=0\ \ (\text{geom. w/ }|r|<1)$$

Therefore, using the sandwich theorem:
$$\lim_{n\to\infty}\frac{\sin^{2n}n}{8^n}=0\text{ and the sequence converges to 0.}$$

Chapter 11: Infinite Series

Terminology:

An **infinite series** (or just **series**) is a sum of all of the terms of an infinite sequence:

$$\sum_{n=1}^{\infty} a_n = S = a_1 + a_2 + \cdots + a_n + \cdots$$

We can also talk about a kth **partial sum,** S_k, which is the sum of the first k terms. For instance, the 3rd partial sum is $a_1 + a_2 + a_3 = S_3$.

We can also use the partial sums to make a **sequence of partial sums,** $\{S_n\}$:

$$\{S_n\} = S_1, S_2, \cdots, S_n, \cdots$$

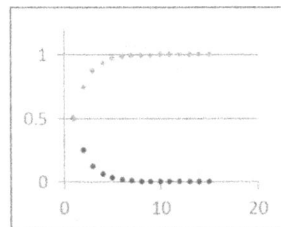

The sequence $\{(1/2)^n\}$ is plotted in dark circles, the series $\sum (1/2)^n$ is plotted in light diamonds

As with sequences, we are interested to find out if a given series **converges**. If it does converge, then it has a finite sum. The formal definition for convergence is:

> **If a series is convergent,**
> **then the sequence of partial sums converges:**
>
> $$\lim_{n \to \infty} S_n = S \text{ and the series sums to } S.$$
>
> **Otherwise, the series diverges and has no sum.**

Determining convergence and sums is the subject of the rest of this chapter as well as later chapters. As we go, we'll talk about special cases since recognizing these special cases helps determine convergence more quickly.

There are many series that are given special names, a few of which are outlined here:

Harmonic Series: The **harmonic series** is the divergent series given by:

$$\sum \frac{1}{n} = 1 + \frac{1}{2} + \frac{1}{3} + \cdots + \frac{1}{n} + \cdots$$

P-Series: The more general **p-series** has the form:

$$\sum \frac{1}{n^p} = 1 + \frac{1}{2^p} + \frac{1}{3^p} + \cdots$$

It converges for $p > 1$, and diverges for $p \leq 1$.

Geometric Series: The **geometric series** is given by:

$$\sum ar^{n-1} = a + ar + ar^2 + \cdots + ar^{n-1} + \cdots$$

where a is a constant. If $|r| < 1$, the series converges and has the sum $\frac{a}{1-r}$. Otherwise it diverges.

Telescoping Series: A **telescoping series** is one where all of the middle terms cancel.

> **Example**
>
> Find the sum: $\sum \left[\frac{1}{n} - \frac{1}{n+1}\right]$:
>
> Writing out terms we get:
> $$\left[\frac{1}{1} - \frac{1}{2}\right] + \left[\frac{1}{2} - \frac{1}{3}\right] + \left[\frac{1}{3} - \frac{1}{4}\right] + \cdots + \left[\frac{1}{n} - \frac{1}{n+1}\right] + \cdots$$
> $$\underbrace{\qquad}_{=\,0} \quad \underbrace{\qquad}_{=\,0}$$
>
> From the pattern, we see that $S_n = 1 - \frac{1}{n+1}$.
> Applying the definitions of convergence and sum:
>
> $$S = \lim_{n \to \infty} (S_n) = \lim_{n \to \infty} \left(1 - \frac{1}{n+1}\right) = 1$$

Tests for Convergence:

There are several general tests for convergence or divergence as outlined below.

The first, the **nth term test**, is often misused and is comprised of two theorems:

> If $\sum a_n$ converges, then $\lim\limits_{n\to\infty} a_n = 0$.
>
> If $\lim\limits_{n\to\infty} a_n \neq 0$, then $\sum a_n$ diverges.

The subtlety that trips people up is assuming that if the limit of the nth term is 0, then the series converges. This is not true (consider the harmonic series). The first theorem predicts the limit *after* convergence is determined. The only definite statement you can make solely from the limit of the nth term is that if it's not zero, the series diverges.

The second method, the **latter terms test**, states that:

> **If the latter terms of two series are identical, then both series diverge or both series converge.**

The idea here is that summing the first finite number of terms gives a non-infinite sum – it's the infinite number of terms that follow that are the real concern.

> *Example*
>
> Does this series converge? $\sum \left[\dfrac{1}{n+1} - \dfrac{1}{n+2}\right]$
>
> Writing out terms we get:
>
> $$\left[\frac{1}{2} - \frac{1}{3}\right] + \left[\frac{1}{3} - \frac{1}{4}\right] + \cdots + \left[\frac{1}{(n-1)+1} - \frac{1}{(n-1)+2}\right] + \cdots$$
>
> This series has the same terms as the telescoping series we saw in the prior example except for the first term. This series must also converge.

Finally, the **properties of addition test** states:

If $\sum a_n = S_A$ and $\sum b_n = S_B$ then

$$\sum(a_n \pm b_n) = S_A \pm S_B \text{ and}$$

$$\sum c a_n = c S_A \text{ and}$$

If $S_B = \infty$, then $\sum(a_n + b_n)$ diverges.

The properties of addition let you combine different series or multiple by a constant. Obviously, adding an infinite sum to a finite one gives another infinite sum as stated by the last part.

Example

Determine if the series converges , and find the sum if it exists:

$$\sum \left(\frac{1}{n}\right)^2 + \left(\frac{1}{2}\right)^n$$

We can split the series into two parts:

1. The first part, $\sum \left(\frac{1}{n}\right)^2$, is a p-series with $p = 2$. According to the definition, this series diverges.

2. No matter what we add to the series, it will diverge. Therefore, the series $\sum \left(\frac{1}{n}\right)^2 + \left(\frac{1}{2}\right)^n$ diverges.

Example

Determine if the series converges, and find the sum if it exists:

$$\Sigma \left(\frac{1}{2}\right)^n$$

Writing out the first terms of this series, we get:

$$\Sigma \left(\frac{1}{2}\right)^n = \frac{1}{2} + \frac{1}{2^2} + \frac{1}{2^3} + \frac{1}{2^4} + \cdots$$

It looks like a geometric series:

$$\Sigma \left(\frac{1}{2}\right)^{n-1} = 1 + \frac{1}{2} + \frac{1}{2^2} + \frac{1}{2^3} + \frac{1}{2^4} + \cdots$$

With $a = 1$; $r = 1/2$. The geometric series would converge to:

$$\frac{a}{1-r} = \frac{1}{1-1/2} = 2.$$

By the latter terms theory, since both series' latter terms are identical, they both converge. And, by the properties of addition:

$$\Sigma \left(\frac{1}{2}\right)^n = -1 + \Sigma \left(\frac{1}{2}\right)^{n-1} = -1 + 2 = 1.$$

Chapter 12: Positive Term Series

Terminology:

A **positive term series** is one in which all terms are greater than zero. In math parlance: $\sum a_n \ni a_n > 0 \; \forall \, n$.

Tests for Convergence:

There are several tests for convergence for positive term series. *Make sure you have a positive term series before applying these tests!*

The first is the **bounded test**. It states that if all of the partial sums are bounded, then so is the series. This reads more like a definition than a useful test.

> If $S_n \leq M \; \forall \, n$, then $\sum a_n$ converges and has sum, $S \leq M$.

The **integral test** is next:

> If $f(x)$ is continuous, decreasing, and positive valued for all $x \geq 1$, then $f(1) + f(2) + \cdots + f(n) + \cdots$ is:
>
> Convergent if $\lim\limits_{t\to\infty} \int_1^t f(x)dx$ converges;
>
> Divergent if $\lim\limits_{t\to\infty} \int_1^t f(x)dx$ diverges.

Example

Show that the harmonic series is divergent:

$$\sum \frac{1}{n} = 1 + \frac{1}{2} + \frac{1}{3} + \cdots + \frac{1}{n} + \cdots$$

Each term is >0, so let $f(x) = \frac{1}{x}$:

$$\lim_{t\to\infty} \int_1^t f(x)dx = \lim_{t\to\infty} \int_1^t \frac{1}{x}dx = \lim_{t\to\infty} (\ln t - \ln 1)$$

$= \infty;$ Thus, the series is divergent.

The **p-series test**, which you may recall from Chapter 11, considers a special subset of positive term series:

$$\sum \frac{1}{n^p} = 1 + \frac{1}{2^p} + \frac{1}{3^p} + \cdots$$

converges for $p > 1$, and diverges for $p \leq 1$.

The **basic comparison test** is:

If both $\sum a_n$ and $\sum b_n$ are positive term series, then

If $\sum b_n$ converges and $a_n \leq b_n \ \forall \ n$, then a_n converges;

If $\sum b_n$ diverges and $a_n \geq b_n \ \forall \ n$, then a_n diverges.

In other words, if a series has a convergent series as an upper bound, then it must converge. If it has a divergent series as a lower bound, it must diverge.

Example

Determine if the series converges or diverges:

$$\sum \frac{1}{3+n^2};$$

We know that $3 + n^2 > n^2 \ \forall \ n \geq 1$

That means that $\frac{1}{3+n^2} < \frac{1}{n^2}$.

The term on the right is a p-series with $p = 2$ which converges. Since our series is upper bounded by the p-series, then it, too, must converge via the basic comparison test.

The **limit comparison test** is another way to compare an unknown series against a known one.

If $\lim\limits_{n\to\infty} \frac{a_n}{b_n} = L > 0$,

then either both series diverge or both converge.

The final two methods get away from the idea of bounding and comparing. The **ratio test** is:

> If $\lim\limits_{n\to\infty} \dfrac{a_{n+1}}{a_n} = L$, then
>
> If $L < 1$, $\sum a_n$ **converges**;
> If $L > 1$, $\sum a_n$ **diverges**.
> If $L = 1$, **the test is inconclusive**.

Similarly, the **root test** states:

> If $\lim\limits_{n\to\infty} \sqrt[n]{a_n} = L$, then
>
> If $L < 1$, $\sum a_n$ **converges**;
> If $L > 1$, $\sum a_n$ **diverges**.
> If $L = 1$, **the test is inconclusive**.

Example

Determine if the series converges or diverges:

$$\sum \frac{2^n}{n^2}.$$

The nth term is: $a_n = \dfrac{2^n}{n^2}$.

The $(n+1)^{\text{th}}$ term is: $a_{n+1} = \dfrac{2^{n+1}}{(n+1)^2}$.

$$\lim_{n\to\infty} \frac{a_{n+1}}{a_n} = \lim_{n\to\infty} \frac{n^2 2^{n+1}}{(n+1)^2 2^n} = \lim_{n\to\infty} \frac{n^2 2}{(n+1)^2}$$

$$= \lim_{n\to\infty} \frac{4n}{2(n+1)} = \lim_{n\to\infty} \frac{4}{2} = 2 > 1.$$

By the ratio test, the series diverges.

Example

Determine if the series converges or diverges:
$$\sum \frac{2^n}{n!}.$$

The nth term is: $a_n = \frac{2^n}{n!}$.

The $(n+1)^{\text{th}}$ term is: $a_{n+1} = \frac{2^{n+1}}{(n+1)!}$.

$$\lim_{n \to \infty} \frac{a_{n+1}}{a_n} = \lim_{n \to \infty} \frac{n! \, 2^{n+1}}{(n+1)! \, 2^n} = \lim_{n \to \infty} \frac{2}{n+1} = 0 < 1$$

By the ratio test, the series converges.

Example

Show that the series $\sum r^n$ converges for $|r| < 1$, and diverges for $|r| > 1$.

The nth root of the nth term is: $\sqrt[n]{a_n} = \sqrt[n]{r^n} = r$.

$$\lim_{n \to \infty} \sqrt[n]{a_n} = \lim_{n \to \infty} r = r$$

By the root test, the series converges for $|r| < 1$, and diverges for $|r| > 1$.

Example

Determine if the series converges or diverges:
$$\sum \frac{n^3 + 3n}{2n + 1}.$$

We know that the series $\sum n^2$ is a positive term series that diverges (it's a p-series with $p = -2$.)

Using the limit comparison test with $b_n = n^2$:

$$\lim_{n \to \infty} \frac{a_n}{b_n} = \lim_{n \to \infty} \frac{n + 3n^{-1}}{2n + 1} = \frac{1}{2} > 0$$

Therefore, the series must also diverge.

Tip: Choose for b_n the highest order of the numerator divided by the highest order of the denominator:

$$\frac{n^3}{n} = n^2$$

Chapter 13: Alternating Series

Terminology:

An **alternating series** is one where the terms change sign from one to the next: $\sum(-1)^n a_n$ or $\sum(-1)^{n-1} a_n$.

To deal with alternating series, you can determine whether the series has convergence or **absolute convergence**.

> **A series is absolutely convergent if**
>
> $$\sum |a_n| = |a_1| + |a_2| + |a_3| + \cdots + |a_n| + \cdots \text{ converges.}$$

Obviously, a convergent positive term series is absolutely convergent.

A **conditionally convergent** series is convergent but not absolutely convergent. In other words, $\sum a_n$ converges but $\sum |a_n|$ diverges

Tests for Convergence:

There are only two tests for convergence on an alternating series. The first one is the **alternating series test**:

> $$\sum(-1)^n a_n \text{ or } \sum(-1)^{n-1} a_n \text{ converges if}$$
>
> $$a_n \geq a_{n+1} > 0 \; \forall \, n \text{ and } \lim_{n \to \infty} a_n = 0,$$

The final one is the **absolute convergence test**:

> **If a series is absolutely convergent,**
> **then it is also convergent.**

In other words, if you turn the alternating series into a positive series and it passes any of the positive series tests, then the original series must converge.

Determine if the series converges or diverges:

$$\sum \frac{(-1)^n}{n}.$$

Taking the limit of a_n: $\displaystyle\lim_{n\to\infty} \frac{1}{n} = 0$

We know that for any $n \geq 1$: $n < n + 1$

Therefore, $\frac{1}{n} > \frac{1}{n+1}$ which means

Example

$a_n \geq a_{n+1} > 0 \; \forall \, n$ and the series converges. Note that this is a conditionally convergent series as $\sum |a_n| = \sum \frac{1}{n}$ is the divergent harmonic series.

Tip: You can also determine if the series is decreasing by checking that $f'(n) < 0$:

$$D_n \left[\frac{1}{n}\right] = \frac{-1}{n^2} < 0$$

Determine if the series converges or diverges:

Example

$$\sum (-1)^n \frac{2^n}{n!}.$$

Rather than trying to prove that $\displaystyle\lim_{n\to\infty} \frac{2^n}{n!} = 0$ and that $a_n \geq a_{n+1} > 0 \; \forall \, n$, it would be easier to consider checking for absolute convergence:

$|a_n| = \frac{2^n}{n!}$ converges as per the ratio test as shown in an example in Chapter 12. Therefore, the series is absolutely convergent.

If $\sum |a_n|$ was divergent, you would have to apply the alternating series test to rule out conditional convergence.

Summary of Series Convergence Tests:

You've learned a number of ways to determine series convergence. The following table summarizes these tests and gives tips to help you pick the right one quickly.

Series	Test	Usage
Any Series	**Summation:** Converges if: $\lim\limits_{n\to\infty} S_n = S \neq \pm\infty$	Telescoping series, Series where a formula for S_n can be obtained
	nth Term: Diverges if $\lim\limits_{n\to\infty} a_n \neq 0$	Easy first test
	Latter terms: Compare later terms to known series ; original series behaves as comparison series	Easy comparison to telescoping, geometric, or p-series
	Properties of Addition: $\sum(a_n \pm b_n) = S_A \pm S_B; \ \sum c a_n = c S_A$	Series that can be broken into parts
Positive Term Series	**Integration:** Con(di)verges if $\lim\limits_{t\to\infty} \int_1^t f(x)dx$ con(di)verges	When nth term is easily integrated
	Basic Comparison: Converges if $\sum b_n$ converges and $a_n \leq b_n$; Diverges if $\sum b_n$ diverges and $a_n \geq b_n$	Easy comparison to geometric or p-series
	Limit Comparison: If $\lim\limits_{n\to\infty} \frac{a_n}{b_n} = L > 0$, $\sum a_n$ behaves as $\sum b_n$	Polynomials
	Ratio: If $\lim\limits_{n\to\infty} \frac{a_{n+1}}{a_n} = L, \begin{cases} L < 1, \text{ converges} \\ L > 1, \text{ diverges} \\ L = 1, ??? \end{cases}$	Factorials, powers of n
	Root: If $\lim\limits_{n\to\infty} \sqrt[n]{a_n} = L, \begin{cases} L < 1, \text{ converges} \\ L > 1, \text{ diverges} \\ L = 1, ??? \end{cases}$	Powers of n
Alternating	**Alternating Series:** Converges if $a_n \geq a_{n+1} > 0$ and $\lim\limits_{n\to\infty} a_n = 0$	Alternating series , good follow up to nth term test
	Absolute Convergence: Converges if $\sum|a_n|$ converges	Try first when alternating series test is difficult

Chapter 14: Power Series

Terminology:

A **power series** has the form:

$$\sum_{n=0}^{\infty} a_n x^n = a_0 + a_1 x + a_2 x^2 + \cdots + a_n x^n + \cdots$$

Notice that the first n is equal to 0.

A **power series in** $(x - c)$ takes the general form $\sum a_n (x - c)^n$ where c is a constant.

$$\sum_{n=0}^{\infty} a_n (x - c)^n = a_0 + a_1 (x - c) + a_2 (x - c)^2 + \cdots \\ + a_n (x - c)^n + \cdots$$

The big difference between a power series and the others we've looked at so far is the x term. We've added a variable to the mix. Rather than a sum of numbers, we have a sum of functions.

We can apply the tests for convergence we've learned thus far (the ratio test is the most helpful), except that there may be x terms for which the series is convergent, and there may be x terms for which the series is divergent. Another way of stating this (and more) is:

> **For $\sum a_n x^n$, only one of the following can be true:**
>
> - **The series converges only for $x = 0$;**
> - **The series absolutely converges for all x;**
> - **The series absolutely converges for $|x| < r$ and diverges for $|x| > r$.**

The term r is the **radius of convergence**, and the interval where $-r < x < r$ is the **interval of convergence**.

For a power series in $(x - c)$, the same is true except that x is replaced by $(x - c)$.

Find the interval of convergence for the following:

$$\Sum \left(\frac{1}{2}\right)^n x^n; n = [0, \infty)$$

The series can be written as: $\sum \left(\frac{x}{2}\right)^n$ which is a geometric series with $a = 1$, and $r = x/2$.

A geometric series converges for

$$|r| = \left|\frac{x}{2}\right| < 1 \rightarrow |x| < 2$$

Find the interval of convergence for the following:

$$\sum \left(\frac{1}{2}\right)^n n x^n$$

Using the ratio test with:

$$u_n = \frac{n x^n}{2^n}; \quad u_{n+1} = \frac{(n+1)x^{n+1}}{2^{(n+1)}};$$

$$\frac{u_{n+1}}{u_n} = \frac{(n+1)2^n x^{n+1}}{n \, 2^{(n+1)} x^n} = \frac{(n+1)x}{2n};$$

The series absolutely converges for:

$$\lim_{n \to \infty} \left|\frac{(n+1)x}{2n}\right| = \lim_{n \to \infty} \left|\frac{x}{2}\right| = \left|\frac{x}{2}\right| < 1 \rightarrow |x| < 2$$

With the ratio and root tests, it is important to check what happens when the test is inconclusive, i.e. when the limit equals 1:

$$\left|\frac{x}{2}\right| = 1 \text{ when } |x| = 2; \quad x = \pm 2:$$

$$x = 2: \sum \left(\frac{x}{2}\right)^n n = \sum n = 1 + 2 + 3 + \cdots$$

$$x = -2: \sum \left(\frac{x}{2}\right)^n n = \sum (-1)^n n = -1 + 2 - 3 + \cdots$$

In the cases where $x = \pm 2$, the resulting series $\sum n$ and $\sum (-1)^n n$ both fail the nth term test:
$$\lim_{n \to \infty} n = \infty \neq 0$$

Diverges in both cases (fails n^{th} term test), so the interval of convergence is $|x| < 2$.

Representations of Functions:

Power series can be used to represent functions over the interval of convergence. A **power series representation of** $f(x)$ is:

$$f(x) = \sum_{n=0}^{\infty} a_n (x - c)^n.$$

Within the radius of convergence, $f(x)$ has a derivative and an integral given by:

$$\frac{d}{dx} f(x) = \sum_{n=0}^{\infty} D_x[a_n (x - c)^n] = \sum_{n=1}^{\infty} n a_n (x - c)^{n-1};$$

$$\int_0^{x-c} f(t) dt = \sum_{n=0}^{\infty} \int_0^{x-c} a_n t^n \ dt = \sum_{n=0}^{\infty} \frac{a_n}{n+1} (x - c)^{n+1}.$$

Note the change in the parameters of the summations, from a lower bound of 0 to 1, in the case of the derivative. Also, in the integral formula a **substitution variable**, t, was used in the integrand so that the independent variable, x, could be used in the upper bound without confusion.

Though some of this can seem abstract at first, we'll consider a few examples that will show how to use the power series representation and its properties. Mostly it's just a matter of pattern recognition. When you get the hang of it, these representations can be useful for approximating functions numerically in software algorithms.

Let's start by considering the series (recall $n = [0, \infty)$ for a power series):
$$\sum 2x^n = 2 + 2x + 2x^2 + 2x^3 + \cdots$$

This is a geometric series with sum:
$$\sum 2x^n = \frac{a}{1-r} = \frac{2}{1-x} \text{ for } |x| < 1.$$

The usual geometric series is $\sum ar^{n-1}, n = [1, \infty)$, so these are equivalent.

As a result, we can say that the function $f(x) = \frac{2}{1-x}$ is represented by the power series $\sum 2x^{n-1}$ when $|x| < 1$. We'll use this result in the following examples.

Example

Find a power series representation of
$$f(x) = \frac{-2}{(1-x)^2}$$

The function above is $D_x\left(\frac{2}{1-x}\right)$.

If $\frac{2}{1-x} = \sum 2x^n$ with $a_n = 2$ and $c = 0$, then

$$\frac{-2}{(1-x)^2} = \sum_{n=0}^{\infty} D_x[2\ x^n] = \sum_{n=1}^{\infty} 2n\, x^{n-1}.$$

Example

Find a power series representation of
$$f(x) = -2 \ln(1-x)$$

The function above is $\int \frac{2}{1-x}\, dx$.

If $\frac{2}{1-x} = \sum 2x^n$ with $a_n = 2$ and $c = 0$, then

$$-2 \ln(1-x) = \sum \int_0^x 2t^n\ dt = \sum \frac{2}{n+1} x^{n+1}.$$

Example

Find a power series representation of
$$f(x) = \tan^{-1} x$$

Recall from Calculus 1 (or integral tables) that
$$\tan^{-1} x = \int \frac{1}{1+x^2}\, dx.$$

The term $\frac{1}{1+x^2}$ is a sum for a geometric series with $a = 1$ and $r = -x^2$. The function $\frac{1}{1+t^2}$ is then represented by the series :

$$\sum(-t^2)^n = \sum(-1)^n t^{2n}.$$

It then follows that
$$\tan^{-1} x = \sum \int_0^x (-1)^n t^{2n} dt = \sum(-1)^n \left(\frac{x^{2n+1}}{2n+1}\right).$$

Chapter 15: Other Series Representations

Taylor & Maclaurin Series:

Here's some convoluted thinking for you. Let's take the general case of $f(x) = \sum_{n=0}^{\infty} a_n (x-c)^n$, do a bunch of derivatives, and then set $x = c$ (and recalling that $0^0 = 1$ and $0! = 1$):

$$f^{(0)}(c) = \sum_{n=0}^{\infty} a_n (x-c)^n \big|_{x=c} = 0!\, a_0;$$

$$\sum_{n=0}^{\infty} a_n (x-c)^n \big|_{x=c}$$
$$= a_0 0^0 + a_1 0^1 + a_2 0^2 + \cdots$$
The only non-zero term is a_0.

$$f^{(1)}(c) = \sum_{n=1}^{\infty} n a_n (x-c)^{n-1} \big|_{x=c} = 1 \cdot a_1 = 1!\, a_1;$$

$$\sum_{n=1}^{\infty} n a_n (x-c)^{n-1} \big|_{x=c}$$
$$= 1 a_1 0^0 + 2 a_2 0^1 + 3 a_3 0^2 + \cdots$$
The only non-zero term is a_1.

$$f^{(2)}(c) = \sum_{n=2}^{\infty} (n-1) n a_n (x-c)^{n-2} \big|_{x=c}$$
$$= 1 \cdot 2 \cdot a_2 = 2!\, a_2;$$

$$\sum_{n=2}^{\infty} (n-1) n a_n (x-c)^{n-2} \big|_{x=c}$$
$$= 1 \cdot 2 \cdot a_2 0^0 + 2 \cdot 3 \cdot a_3 0^1$$
$$+ 3 \cdot 4 \cdot a_3 0^2 + \cdots$$
The only non-zero term is $2a_2$.

$$f^{(3)}(c) = \sum_{n=3}^{\infty} (n-2)(n-1) n a_n (x-c)^{n-2} \big|_{x=c}$$
$$= 1 \cdot 2 \cdot 3 \cdot a_3 = 3!\, a_3;$$

$$\vdots$$

$$f^{(k)}(c) =$$
$$\sum_{n=k}^{\infty} (n - (k-1)) \cdots (n-1) n a_n (x-c)^{n-1} \big|_{x=c}$$
$$= 1 \cdot 2 \cdot 3 \cdots \cdot (k-1) \cdot k \cdot a_k = k!\, a_k.$$

Noting the pattern, in general we can say that

$$f^{(n)}(c) = n!\, a_n \;\rightarrow\; a_n = \frac{f^{(n)}(c)}{n!}.$$

Now we can give the **Taylor series representation of $f(x)$**:

> If $f(x) = \sum_{n=0}^{\infty} a_n (x-c)^n$ and $c \in |x| < r$,
>
> Then $f(x) = \sum_{n=0}^{\infty} \frac{f^{(n)}(c)}{n!} (x-c)^n$.

The condition $c \in |x| < r$, means that c is in the interval of convergence.

The **Maclaurin series representation of $f(x)$** is just the Taylor series with $c = 0$:

> **If $f(x) = \sum_{n=0}^{\infty} a_n x^n$ and $|x| < r$,**
>
> **Then $f(x) = \sum_{n=0}^{\infty} \frac{f^{(n)}(0)}{n!} x^n$.**

In the long run, series representation such as these can help with more complex integration problems. Whether you fully understand the above derivation is much less important that being able to use the series representations.

> **Example**
>
> Find a series representation of e^x.
>
> Since $c = 0$, we'll use the Maclaurin series:
>
> $$f(x) = \sum_{n=0}^{\infty} \frac{f^{(n)}(0)}{n!} x^n$$
>
> The derivative of e^x is equal to e^x :
>
> $$f^{(0)}(0) = f^{(1)}(0) = f^{(2)}(0) = \cdots$$
>
> $$= f^{(n)}(0) = e^0 = 1$$
>
> Plugging in:
>
> $$e^x = \sum_{n=0}^{\infty} \frac{1}{n!} x^n = 1 + x + \frac{x^2}{2!} + \frac{x^3}{3!} + \frac{x^4}{4!} + \cdots$$
>
> Note: Because $\sum \frac{x^n}{n!}$ converges for all x (see Problem 14.3), the interval of convergence is $(-\infty, \infty)$

This is an important result with a lot of uses…

> **Example**
>
> Evaluate $\int x e^x dx$.
>
> Using the series representation for e^x to simplify the problem and using the first few terms (more terms gives more accuracy):
>
> $$\int x e^x dx = \int x \left[1 + x + \frac{x^2}{2} \right] dx = \frac{x^2}{2} + \frac{x^3}{3} + \frac{x^4}{8} + c$$

Find a series representation of $\sin x$.

Example

Since $c = 0$, we'll use the Maclaurin series:

$$f(x) = \sum_{n=0}^{\infty} \frac{f^{(n)}(0)}{n!} x^n$$

Taking the derivatives of $\sin x$ and setting $x = 0$:

$f^{(0)}(0) = \sin(0) = 0;$

$f^{(1)}(0) = \cos(0) = 1;$

$f^{(2)}(0) = -\sin(0) = 0;$

$f^{(3)}(0) = -\cos(0) = -1; \cdots$

The following pattern emerges:

$$f^{(2n)}(0) = 0; \quad f^{(2n+1)}(0) = (-1)^n;$$

Only the odd terms are non-zero.

Plugging in:

$$\sin x = \sum_{n=0}^{\infty} (-1)^n \frac{x^{2n+1}}{(2n+1)!}$$

$$= x - \frac{x^3}{3!} + \frac{x^5}{5!} - \frac{x^7}{7!} + \cdots$$

Note: Because $\sum \frac{x^n}{n!}$ converges for all x (see Problem 14.3), the interval of convergence is $(-\infty, \infty)$

Binomial Series:

The last functional series representation of this course is developed with the use of the **Binomial Theorem** for $(a + b)^k$ with $a = 1$ and $b = x$. The **Binomial Series** is used to represent the function $(1 + x)^k$ as follows:

If $|x| < 1$, then for any real number, k:

$$(1 + x)^k = 1 + kx + \frac{k(k - 1)}{2!} x^2 + \cdots$$
$$+ \frac{k(k - 1) \cdots (k - n + 1)}{n!} x^n + \cdots$$

Let's see the Binomial Series in action.

Example

Find a series representation of $\sqrt[4]{1 + x}$.

$$(1 + x)^{1/4} = 1 + \frac{x}{4} + \frac{\left(\frac{1}{4}\right)\left(-\frac{3}{4}\right)}{2!} x^2 + \frac{\left(\frac{1}{4}\right)\left(-\frac{3}{4}\right)\left(-\frac{7}{4}\right)}{3!} x^3 \cdots$$

$$= 1 + 0.25x - 0.094x^2 + 0.055x^3 + \cdots$$

Example

Approximate $\int_0^{0.2} \sqrt[4]{1 + x^2} \, dx$.

First we find a series representation:

$$(1 + x^2)^{1/4} = 1 + \frac{x^2}{4} + \frac{\left(\frac{1}{4}\right)\left(-\frac{3}{4}\right)}{2!} x^4 + \frac{\left(\frac{1}{4}\right)\left(-\frac{3}{4}\right)\left(-\frac{7}{4}\right)}{3!} x^6 \cdots$$

$$\cong 1 + 0.25x^2 - 0.094x^4$$

Plugging into the integral :

$$\int_0^{0.2} \sqrt[4]{1 + x^2} \, dx \cong x + \frac{0.25x^3}{3} - \frac{0.094x^5}{5} \Big|_0^{0.2}$$

$$= 0.2 + 0.00067 - 6.0 \times 10^{-6} = 0.2007$$

Accuracy of Series Representations:

As our final topic, we consider **accuracy**. If you are representing a function with an infinite series, by hand or with a computer algorithm, you would be wise to ask how many terms are needed to get a reasonably accurate answer. For alternating series (see Chapter 13) with a few conditions, we have the following theorem to estimate **error**:

> **If an infinite series is:**
> - **an alternating series**
>   - $\sum_{n=1}^{\infty}(-1)^{n-1}a_n$
> - **with decreasing absolute values in each successive term,**
>   - $a_n > a_{n+1}$
> - **and the nth term approaches 0**
>   - $\lim_{n\to\infty} a_n = 0$
>
> **Then:**
> - **the sum can be expressed as the partial sum of n terms, S_n, with an error, E, less than the value of the next term, a_{n+1}.**
>   - $S \sim S_n$ with $E < a_{n+1}$

Example

Approximate $\sin(0.2)$ to four decimal places:

We know from a prior example that
$$\sin x = x - \frac{x^3}{3!} + \frac{x^5}{5!} - \frac{x^7}{7!} + \cdots$$

Plugging in $x = 0.2$ we get the table to the right.

We only need the first two terms, and the error is 2.6×10^{-6}. Compare to a calculator result:
$$\sin(0.2) = 0.198669 \ldots$$

n	0	1	2
a_n	0.2	-0.0013	2.6×10^{-6}
$\sum a_n$	0.2	0.1987	0.1987

<u>Appendices</u>

A: Course Summary
B: Problem Sets
C: Solutions to Problem Sets
D: Derivative Tables
E: Integral Tables

Appendix A: Course Summary

1 **Multi-variable functions** have
independent variables that make up the **domain** and
dependent variables that make up the **range**.
A **discontinuity**: limit to a point from different directions yields different answers.

2 **Partial derivatives**: Same as before, pick a variable, treat rest as constants.
3 Obey same properties as full derivatives.

Chain Rule	Implicit Chain Rule	Increments
$\dfrac{\partial w}{\partial x} = \dfrac{\partial w}{\partial u} \cdot \dfrac{\partial u}{\partial x} + \dfrac{\partial w}{\partial v} \cdot \dfrac{\partial v}{\partial x}$	$\dfrac{dy}{dx} = -\dfrac{\partial f/\partial x}{\partial f/\partial y}$	$dw = \dfrac{\partial w}{\partial x} dx + \dfrac{\partial w}{\partial y} dy$

4 **Max/Min**:

Critical point: $\quad \dfrac{\partial f}{\partial x} = \dfrac{\partial f}{\partial y} = 0;$ OR $\dfrac{\partial f}{\partial x}$ or $\dfrac{\partial f}{\partial y}$ DNE

Max or Min: $\quad f_{xx} < 0;\ f_{xx}f_{yy} - f_{xy}^2 > 0\ \to$ max;
$\quad\quad\quad\quad\quad\quad f_{xx} > 0;\ f_{xx}f_{yy} - f_{xy}^2 > 0\ \to$ min.

5 **Partial integration**: Same as before, pick a variable, treat rest as constants.
Obey same properties as full integration.

Cartesian	6 Cylindrical	7 Spherical
$\dfrac{\partial}{\partial x}, \dfrac{\partial}{\partial y}, \dfrac{\partial}{\partial z}$	$\dfrac{\partial}{\partial r}, \dfrac{1}{r}\dfrac{\partial}{\partial \theta}, \dfrac{\partial}{\partial z}$	$\dfrac{\partial}{\partial \rho}, \dfrac{1}{\rho}\dfrac{\partial}{\partial \varphi}, \dfrac{1}{\rho \sin \varphi}\dfrac{\partial}{\partial \theta}$
dx, dy, dz	$dr, r d\theta, dz$	$d\rho, \rho d\varphi, \rho \sin \varphi\, d\theta$
$x = r \cos \theta$ $y = r \sin \theta$ $z = z$ or $\rho \cos \varphi$	$r = \sqrt{x^2 + y^2}$ $\theta = \theta$ or $\tan^{-1}\left(\frac{y}{x}\right)$ $z = z$ or $\rho \cos \varphi$	$\rho = \sqrt{r^2 + z^2}$ $\varphi = \tan^{-1}\left(\frac{r}{z}\right)$ $\theta = \theta$ or $\tan^{-1}\left(\frac{y}{x}\right)$

8 Length, Area, Volume: $\quad A_{cross-section} = \int f(x,y)dx\,;V = \iint f(x,y)dxdy\,.$

$$L = \int dL\,;\quad A = \iint dA\,;\quad V = \iiint dV.$$

9 **Limits** with indeterminate forms:

$$\frac{g(a)}{h(a)} = \frac{0}{0} \, or \, \frac{\infty}{\infty} \rightarrow L = \lim_{x \to a} \frac{g'(x)}{h'(x)} \quad \text{(L'Hôptial's Rule)}$$

$$g(a)h(a) = 0 \cdot \infty \rightarrow \text{invert one function, use L'Hôptial's Rule}$$

$$g(a)^{h(a)} = 0^0, \infty^0, \text{ or } 1^\infty \rightarrow L_{new} = \lim_{x \to a}[h(x) \ln g(x)], \; L = e^{L_{new}}$$

10 Infinite Sequences: Define by **nth term**; Define **recursively** (a_1 and $a_k = f(a_{k-1})$).
   Sequence Convergence:
   Limit; Sandwich Theorem; Bounded Monotonic; Geometric

Infinite Series: $\sum_{n=1}^{\infty} a_n = S$
   Series Convergence: (See page 63 for details)

(See page 63 for details)

11-13

Any Series	Positive Term	Alternating
Summation	Integration	Alternating Series
nth Term	Basic Comparison	Absolute Convergence
Latter terms	Limit Comparison	
Properties of Addition	Ratio	
	Root	

14 Power Series: $\sum a_n(x - c)^n$

Representations of Functions:
$$f(x) = \sum_{n=0}^{\infty} a_n (x - c)^n.$$

$$\frac{d}{dx}f(x) = \sum_{n=1}^{\infty} na_n (x - c)^{n-1}; \quad \int_0^{x-c} f(t)dt = \sum_{n=0}^{\infty} \frac{a_n}{n+1}(x - c)^{n+1}$$

15 $$f(x) = \sum_{n=0}^{\infty} \frac{f^{(n)}(c)}{n!}(x - c)^n \quad \text{(Taylor \& Maclaurin } (c = 0))$$

$$(1 + x)^k = 1 + kx + \frac{k(k-1)}{2!}x^2 + \cdots + \frac{k(k-1)\cdots(k-n+1)}{n!}x^n \cdots \quad \text{(Binomial)}$$

Alternating Series Estimation Error: $S \sim S_n$ with $E < a_{n+1}$

Chapter correlations are given in large grey font.

Summary of Special Sequences and Series

Type	Name	Comments
Sequences:	Arithmetic	Add/subtract by constant
	Geometric	Multiply/divide by constant
	Alternating	Change signs
Series:	Harmonic	$\sum \frac{1}{n}$
	P-Series	$\sum \frac{1}{n^p}$; converges $p > 1$
	Geometric	$\sum ar^{n-1}$; $\|r\| < 1 \to s = \frac{a}{1-r}$
	Telescoping	Middle terms cancel; $S = \lim_{n \to \infty} (S_n)$
	Power Series	$\sum_{n=0}^{\infty} a_n (x - c)^n$
	Taylor	$\sum_{n=0}^{\infty} \frac{f^{(n)}(c)}{n!} (x - c)^n.$
	Maclaurin	$\sum_{n=0}^{\infty} \frac{f^{(n)}(0)}{n!} x^n$
	Binomial	$(1 + x)^k$

Appendix B: Problem Sets

Chapter 1

Determine if the following functions are discontinuous at (0,0):

1.1: $\dfrac{y}{x+1}$

1.2: $\dfrac{4x+1}{2-xy}$

1.3: $\dfrac{2x+3y}{x-y}$

1.4: $\dfrac{x^2}{xy}$

1.5: $\dfrac{xy}{xy-2}$

1.6: $\dfrac{x+2y+1}{x-2y+2}$

1.7: $\dfrac{x+2y}{x+y}$

1.8: $\dfrac{x^2-y^2}{x^2+xy-1}$

1.9: $\dfrac{x^2-y^2}{x^2+2}$

1.10: $\dfrac{x^2+y^2}{3x^2+2xy}$

Find where the following functions are discontinuous and prove they are discontinuity at one point (x, y):

1.11: $\dfrac{4x}{x+y}$

1.12: $\dfrac{5xy}{x^2-y^2}$

1.13: $\dfrac{x}{3-y}$

1.14: $\dfrac{xy}{y^2-1}$

Chapter 2

Find the second partial derivatives for the following:

2.1: $x + 2y$

2.2: $x^2 + 4xy + 3y^2$

2.3: $x^2 + yx$

2.4: $\sqrt{x^2 + y^2}$

2.5: $\sin x + \cos y$

2.6: $y^2 \sin 3x + x^2 \cos 4y$

2.7: e^{xy}

2.8: $\ln(x^2 + y^2)$

2.9: $xy \cos 2x$

2.10: $\tan^{-1}(x^2 + y^2)$

Find $\dfrac{\partial^3 f}{\partial x^3}$, $\dfrac{\partial^3 f}{\partial y^3}$, and $\dfrac{\partial^3 f}{\partial x^2 \partial y}$ for the following:

2.11: $4x^3 y + 3y^2$

2.12: $(x + 1)^3 yz + (2y - 1)^2 xz$

2.13: $\sin xy$

2.14: $e^{x^2 - y^2}$

Chapter 3

Find the change in w near the point $(1,1)$ if $dx = dy = 0.01$:

3.1: $w = x^2y - y^2x$ **3.2:** $w = 4x^2 + 2xy + y^2$

3.3: $w = \frac{xy}{x+y}$ **3.4:** $w = y \sin(1 - x)$

3.5: $w = e^{x^2y}$ **3.6:** $w = \ln(2x - y)$

Evaluate $\frac{\partial w}{\partial x}$ directly and by using the Chain Rule::

3.7: $u = xy; \ v = x^2y^2;$ **3.8:** $u = x - y; \ v = \ln x;$
$\quad\quad w = 2u - 3v$ $\quad\quad w = uv - u^2$

3.9: $u = x^2; \ v = 2x + y;$ **3.10:** $u = \cos 2x; \ v = \sin y;$
$\quad\quad w = u/v$ $\quad\quad w = uv$

3.11: $u = e^{xy}; \ v = x + y;$ **3.12:** $u = xy; \ v = x - y;$
$\quad\quad w = 4u^2 + v$ $\quad\quad w = \tan^{-1}(2u + v)$

Assume $f(x, y) = 0$ implicitly determines y and find y':

3.13: $x^4y = y^3x$ **3.14:** $x^2y^2 = xy + \cos x$

Chapter 4

Find the critical points and determine if they are maxima or minima:

4.1: $x^2 - 2xy + 4y^2$ **4.2:** $x^2y - y^2x$

4.3: $(x-1)^2 + (y-2)^2$ **4.4:** $y^3 - x^2y$

4.5: $(4x+2)^2 + y^2$ **4.6:** $e^{x^2+y^2}$

4.7: $\cos^2 x + \cos^2 y$ **4.8:** $x^2y^2 - 2xy + (y-1)^2$

4.9: $x^3 + 3xy - y^3$ **4.10:** $x^3y - 3xy - y^2$

4.11: $x^3 - 3x^2y + y^4$ **4.12:** $x^3 - 3xy + xy^3$

4.13: $x^2y^2 - 2xy$ **4.14:** $x^4y^2 - x^2 - y^2$

Chapter 5

Evaluate $\iint f(x,y)dxdy$ and $\iint f(x,y)dydx$:

5.1: $\int_0^3 \int_0^2 x^2 + y^2 \, dx \, dy$

5.2: $\int_1^2 \int_0^2 4xy + 3x^2 \, dx \, dy$

5.3: $\int_0^1 \int_0^1 3x^2 y + 2(x-1)y^2 \, dxdy$

5.4: $\int_0^{\pi/2} \int_0^\pi \sin x \cos y \, dx \, dy$

5.5: $\int_0^3 \int_0^1 y^2 e^{2x} \, dx \, dy$

5.6: $\int_0^1 \int_0^\pi y \sin x \, dx \, dy$

5.7: $\int_0^1 \int_1^2 (2x-4)^2 (4y+2) \, dx \, dy$

5.8: $\int_0^1 \int_0^1 x(xy-1)^2 \, dx \, dy$

Integrate the function $f(x,y)$ in the region specified:

5.9: $f(x,y) = x^3 + yx$;
$y = x$; $y = x^2$; $0 \le x \le 1$

5.10: Reverse the order of integration for Problem 5.9

5.11: $f(x,y) = xy$;
$x = 0$; $x = 2y$; $0 \le y \le 1$

5.12: Reverse the order of integration for Problem 5.11

5.13: $f(x,y) = x^2 y$;
$y = x$; $y = 2x$; $0 \le x \le 2$

5.14: Reverse the order of integration for Problem 5.13

Chapter 6

Convert the following Cartesian expressions to Cylindrical coordinates:

6.1: $(4, 5, -2)$ **6.2:** $(7, 2, 12)$

6.3: $f(x, y, z) = 4x + y^2 z$ **6.4:** $f(x, y, z) = z(x^2 + y^2)$

Find the derivatives with respect to r and θ:

6.5: $(4r^2 - 1)^2 \cos \theta$ **6.6:** $\dfrac{r^3}{\sin \theta}$

6.7: $\ln(r^2 \theta)$ **6.8:** $e^{2r \cos \theta}$

Integrate the function over the region specified:

6.9: $f(r, \theta) = r\theta - 1;$
$0 \le r \le 2; \ 0 \le \theta \le \pi$

6.10: $f(r, \theta) = r^2 \theta - 3\theta^2 r;$
$0 \le r \le 1; \ 0 \le \theta \le \pi$

6.11: $f(r, \theta) = \cos r\theta;$
$0 \le r \le 1/2; \ 0 \le \theta \le \pi/2$

6.12: $f(r, \theta) = \sin r^2 \cos \theta;$
$0 \le r \le \sqrt{\pi/2}; \ 0 \le \theta \le \pi/2$

6.13: $f(r, \theta) = 4(r\theta - 1)^3;$
$0 \le r \le 2; \ 0 \le \theta \le \pi$

6.14: $f(r, \theta) = \theta e^{r^2};$
$0 \le r \le 1; \ 0 \le \theta \le \pi$

Chapter 7

Convert the following Cartesian expressions to Cylindrical coordinates:

7.1: $(4, 5, -2)$ **7.2:** $(7, 2, 12)$

7.3: $f(x, y, z) = 4x + y^2 z$ **7.4:** $f(x, y, z) = z(x^2 + y^2)$

Find the derivatives with respect to r, φ, and θ:

7.5: $(4\rho^2 - 1)^2 \cos\theta \sin\varphi$ **7.6:** $\dfrac{\rho^3 \sin\varphi}{\sin\theta}$

7.7: $\ln(\rho^2 \theta \varphi)$ **7.8:** $e^{2\rho \sin\varphi \cos\theta}$

Integrate the function over the region specified:

7.9: $f(\rho, \varphi, \theta) = \rho \cos\varphi - 1$; $0 \le \rho \le 2$; $0 \le \varphi \le \pi/2$

7.10: $f(\rho, \varphi, \theta) = \rho^2 \theta - 3\theta^2 \rho$; $0 \le \rho \le 1$; $0 \le \theta \le \pi$

7.11: $f(\rho, \varphi, \theta) = \cos\rho\varphi$; $0 \le \rho \le 1/2$; $0 \le \varphi \le \pi/2$

7.12: $f(\rho, \varphi, \theta) = \sin\rho^2 \cos\theta$; $0 \le \rho \le \sqrt{\pi/2}$; $0 \le \theta \le \pi/2$

7.13: $f(\rho, \varphi, \theta) = 4(\rho\varphi - 1)^3$; $0 \le \rho \le 2$; $0 \le \varphi \le \pi$

7.14: $f(\rho, \varphi, \theta) = \theta e^{\rho^2 \sin\varphi}$; $0 \le \rho \le 1$; $0 \le \theta \le \pi$

Chapter 8

Find the cross-sectional area of the following:

8.1: $f(x,z) = 3z^3x^2$;
$0 \le x \le 1$; $z = 2$

8.2: $f(x,y) = x \sin xy$;
$0 \le y \le \pi$; $x = 1/2$

8.3: $f(r,z) = 2r \sin rz$;
$0 \le z \le 1$; $r = \pi$

8.4: $f(r,\theta) = 2 \cos r\theta$;
$0 \le \theta \le 1/2$; $r = \pi$

8.5: $f(\rho, \varphi) = \rho^2 e^{\cos \varphi}$;
$0 \le \rho \le 1$; $\varphi = 0$

8.6: $f(\rho, \varphi) = \sin \varphi$;
$0 \le \varphi \le \pi$; $\rho = 10$

Take a triple integral over the volume indicated:

8.7: A cylinder of radius R
and height H

8.8: $f(r,\theta,z) = r\theta z$
Over a cylinder
of radius 2, height 3

8.9: $f(r,\theta,z) = \cos r\theta$
Over a cylinder
of radius 1/2, height 4

8.10: $f(r,\theta,z) = ze^{r\theta}$
Over a cylinder
of radius 1, height 2

8.11: $f(\rho, \varphi, \theta) = \rho^3$
Over a sphere of radius 2

8.12: $f(\rho, \varphi, \theta) = \cos^2 \varphi$
Over a sphere of radius 3

8.13: $f(\rho, \varphi, \theta) = \rho e^{\cos \varphi}$
Over a sphere of radius 1

8.14: $f(\rho, \varphi, \theta) = \dfrac{\ln(\cos \varphi)}{\rho^2}$
Over a sphere of radius 2

Chapter 9

Solve the following:

9.1: $\lim\limits_{x\to 1} \dfrac{x^2-2x+1}{x^3+2x-3}$

9.2: $\lim\limits_{x\to -2} \dfrac{x^2-4}{x^3-4x}$

9.3: $\lim\limits_{x\to 0} \dfrac{10x}{\sin 2x}$

9.4: $\lim\limits_{x\to \infty} \dfrac{5x^2+2}{\ln 2x}$

9.5: $\lim\limits_{x\to \infty} \dfrac{x^4+3x^3-2x^2+x-1}{2x^4+5x^2-2}$

9.6: $\lim\limits_{x\to \infty} x^2 4^{-x}$

9.7: $\lim\limits_{x\to \infty} x \sin\dfrac{1}{x}$

9.8: $\lim\limits_{x\to 0} x \sin x$

9.9: $\lim\limits_{x\to \infty} e^{-x} \ln x$

9.10: $\lim\limits_{x\to 0} x^x$

9.11: $\lim\limits_{x\to 0} (e^x)^{\cot x}$

9.12: $\lim\limits_{x\to 0} (\cos x)^{1/x}$

9.13: $\lim\limits_{x\to \infty} (x)^{e^{-x}}$

9.14: $\lim\limits_{x\to 1} \left(\dfrac{1}{x-1}\right)^{\ln x}$

Chapter 10

Determine if the following sequences converge or diverge. If they converge, find the limit:

10.1: $a_1 = 4$; $a_k = 4 - a_{k-1}$

10.2: $a_1 = 5$; $a_k = a_{k-1} + 3$

10.3: $a_1 = 10$; $a_k = 10a_{k-1}$

10.4: $a_1 = 64$; $a_k = \frac{a_{k-1}}{2}$

10.5: $\left\{\frac{5n^2+2}{\ln 2n}\right\}$

10.6: $\{e^{-n}\ln n\}$

10.7: $\left\{\left(\frac{1}{n}\right)^n\right\}$

10.8: $\{n^{1/n}\}$

10.9: $\left\{2 - \frac{1}{n}\right\}$

10.10: $\left\{\left(1 + \frac{2}{n}\right)^{2n}\right\}$

10.11: $\left\{2 + \frac{1}{2^n}\right\}$

10.12: $\{(-1)^{n+1}10^{-n}\}$

10.13: $\{4^{-n}\cos 4n\}$

10.14: $\left\{\frac{n^3+3}{n^2+4n-2}\right\}$

Chapter 11

Determine if the following series converge or diverge, and find the sum when possible:

11.1: $\sum \frac{4n^2+2n+5}{3n^2-3n+5}$

11.2: $\sum \frac{n^3}{3n^3+2n^2-2n-3}$

11.3: $\sum \left(1+\frac{3}{n}\right)^n$

11.4: $\sum n^{1/2n}$

11.5: $2 + 0.2 + 0.02 + \cdots$

11.6: $0.04 + 0.0004 + 0.000004 + \cdots$

11.7: $\sum \left(\frac{1}{3}\right)^n \left(\frac{1}{4}\right)^{n-1}$

11.8: $3 + \frac{3}{4} + \frac{3}{9} + \frac{3}{16} + \frac{3}{25} + \cdots$

11.9: $5 + \frac{5}{2} + \frac{5}{3} + \frac{5}{4} + \cdots$

11.10: $\sum \left[\frac{1}{2n-1} - \frac{1}{2n+1}\right]$

11.11: $\sum \left[\left(\frac{1}{2}\right)^n + \left(\frac{2}{n}\right)^n\right]$

11.12: $\sum \left[\left(\frac{1}{3}\right)^n + \left(\frac{1}{4}\right)^{n-1}\right]$

11.13: $\sum \frac{3}{n(n+1)}$

11.14: $\sum \frac{1}{n^2+3n+2}$

Chapter 12

Determine if the following series converge or diverge:

12.1: $\sum \dfrac{n^2}{n^3+3}$

12.2: $\sum \dfrac{n^2}{(n^3+3)^2}$

12.3: $\sum n e^{-n^2}$

12.4: $\sum \dfrac{\left|\cos\left(\frac{1}{n^2}\right)\right|}{n}$

12.5: $\sum \dfrac{1}{n2^n}$

12.6: $\sum \dfrac{\cos^2 n}{3^n}$

12.7: $\sum \dfrac{n}{2n^2-1}$

12.8: $\sum \dfrac{2}{2+n^2}$

12.9: $\sum \dfrac{n+1}{2n^2-1}$

12.10: $\sum \dfrac{1}{\sqrt{n}+1}$

12.11: $\sum \dfrac{n^2+2}{n!}$

12.12: $\sum \dfrac{n!}{ne^n}$

12.13: $\sum \dfrac{2^n}{n^2+2}$

12.14: $\sum \dfrac{2^n}{n^{n+2}}$

Chapter 13

Determine if the following are absolutely convergent, conditionally convergent, or divergent:

13.1: $\sum(-1)^n \frac{n}{n^2+1}$

13.2: $\sum(-1)^n \frac{4}{2n^3-1}$

13.3: $\sum(-1)^n \frac{n}{n+1}$

13.4: $\sum\left(\frac{n^3+1}{-n^2}\right)^n$

13.5: $\sum(-1)^n \left(\frac{4}{n}\right)^{1/n}$

13.6: $\sum(-1)^n \frac{3}{1+\sqrt{n}}$

13.7: $\sum(-1)^n \frac{n}{2^n}$

13.8: $\sum(-1)^n \frac{e^n}{n!}$

13.9: $\sum(-1)^n \frac{n^2}{n^3+3}$

13.10: $\sum(-1)^n \frac{1}{\sqrt{n+1}}$

13.11: $\sum(-1)^n \frac{3^{1/n}}{n!}$

13.12: $\sum(-1)^n \frac{4^n}{n^{n+1}}$

13.13: $\sum(-1)^n \frac{1}{n\sqrt{\ln(n)}}$

13.14: $\sum(-1)^n \frac{n^3+3n}{n^2+3n-2}$

Chapter 14

Determine the interval of convergence for the following:

14.1: $\sum \frac{4x^n}{n^2}; n > 0$

14.2: $\sum \frac{nx^n}{2^n}$

14.3: $\sum \frac{x^n}{n!}$

14.4: $\sum \frac{3^{1/n}x^n}{n!}$

14.5: $\sum \frac{(4x)^n}{n^{n+1}}$

14.6: $\sum \frac{nx^n}{n+1}$

Find a power series representation for the following:

14.7: $\frac{2}{1-(x-1)^3}$

14.8: $\frac{6(x-1)^2}{(1-(x-1)^3)^2}$

14.9: $\frac{1}{4+2x}$

14.10: $\frac{-2}{(4+2x)^2}$

14.11: $\frac{1}{2}\ln(4+2x)$

14.12: $\frac{x}{1-x^2}$

14.13: $\frac{1+x^2}{(1-x^2)^2}$

14.14: $-\frac{1}{2}\ln(1-x^2)$

Chapter 15

Find representations or give approximations for the following:

15.1: $\cos x$

15.2: $\ln(1+x)$

15.3: $\tan^{-1} x$

15.4: $\sinh x$

15.5: $\cosh x$

15.6: e^{-x^2}

15.7: $\ln 1.1$

15.8: $\tan^{-1} 0.4$

15.9: $\int_0^{0.1} x \sin x \ dx$

15.10: $\int_0^{0.1} x^2 \tan^{-1} x \ dx$

15.11: $\int_0^{0.1} x^2 \cos 2x \ dx$

15.12: $\int_0^{0.1} x^3 \ln(1-x) \ dx$

15.13: $\int_0^{0.1} \sqrt{1+\sqrt{x}} \ dx$

15.14: $\int_0^{0.1} x\sqrt[4]{1-x} \ dx$

Appendix C: Solutions to Problem Sets

Chapter 1

1.1: $\frac{y}{x+1}$ Let $y = mx$:

$$\lim_{(x,y)\to(0,0)} \frac{y}{x+1} = \lim_{x\to0} \frac{mx}{x+1} = 0$$

Does not depend on m, so continuous at (0,0)

1.2: $\frac{4x+1}{2-xy}$ Let $y = mx$:

$$\lim_{(x,y)\to(0,0)} \frac{4x+1}{2-xy} = \lim_{x\to0} \frac{4x+1}{2-mx^2} = \frac{1}{2}$$

Does not depend on m, so continuous at (0,0)

1.3: $\frac{2x+3y}{x-y}$ Let $y = mx$:

$$\lim_{(x,y)\to(0,0)} \frac{2x+3y}{x-y} = \lim_{x\to0} \frac{2x+3mx}{x-mx}$$

$$= \frac{2+3m}{1-m}$$

Does depend on m: discontinuous

1.4: $\frac{x^2}{xy}$ Let $y = mx$:

$$\lim_{(x,y)\to(0,0)} \frac{x^2}{xy} = \lim_{x\to0} \frac{x^2}{mx^2} = \frac{1}{m}$$

Does depend on m: discontinuous

1.5: $\frac{xy}{xy-2}$ Let $y = mx$:

$$\lim_{(x,y)\to(0,0)} \frac{xy}{xy-2} = \lim_{x\to0} \frac{mx^2}{mx^2-2} = 0$$

Does not depend on m: continuous

1.6: $\frac{x+2y+1}{x-2y+2}$ Let $y = mx$:

$$= \lim_{x\to0} \frac{x+2mx+1}{x-2mx+2} = \frac{1}{2}$$

Does not depend on m: continuous

1.7: $\frac{x+2y}{x+y}$ Let $y = mx$:

$$\lim_{x\to0} \frac{x+2mx}{x+mx} = \frac{1+2m}{1+m}$$

Does depend on m: discontinuous

1.8: $\frac{x^2-y^2}{x^2+xy-1}$ Let $y = mx$:

$$\lim_{x\to0} \frac{x^2-m^2x^2}{x^2+mx^2-1} = 0$$

Does not depend on m: continuous

1.9: $\frac{x^2-y^2}{x^2+2}$ Let $y = mx$:

$$\lim_{x\to0} \frac{x^2-m^2x^2}{x^2+2} = 0$$

Does not depend on m: continuous

1.10: $\frac{x^2+y^2}{3x^2+2xy}$ Let $y = mx$:

$$\lim_{x\to0} \frac{x^2+m^2x^2}{3x^2+2mx^2} = \frac{1+m^2}{3+2m}$$

Does depend on m: discontinuous

1.11: $\dfrac{4x}{x+y}$

Discontinuous when $x = -y$.

Choose $x = y = 0$ and let $y = mx$:

$$\lim_{(x,y)\to(0,0)} \frac{4x}{x+y} = \lim_{x\to 0} \frac{4x}{x+mx} = \frac{4}{1+m}$$

Limit depends on m, so discontinuous at (0,0) by two path rule.

1.12: $\dfrac{5xy}{x^2-y^2}$

Discontinuous when $x^2 = -y^2$.

Choose $x = y = 0$ and let $y = mx$:

$$\lim_{(x,y)\to(0,0)} \frac{5xy}{x^2-y^2} = \lim_{x\to 0} \frac{5mx^2}{x^2-(mx)^2}$$

$$= \frac{5m}{1-m^2}$$

Limit depends on m, so discontinuous at (0,0) by two path rule.

1.13: $\dfrac{x}{3-y}$

Discontinuous when $y = 3$.

Choose $x = 0$; $y = 3$.

An equation for a line going through this point is $y = mx + 3$:

$$\lim_{(x,y)\to(0,3)} \frac{x}{3-y} = \lim_{x\to 0} \frac{x}{3-mx-3}$$

$$= \frac{-1}{m}$$

Limit depends on m, so discontinuous at (0,3) by two path rule.

1.14: $\dfrac{xy}{y^2-1}$

Discontinuous when $y = \pm 1$.

Choose $x = 0$; $y = 1$.

An equation for a line going through this point is $y = mx + 1$:

$$\lim_{(x,y)\to(0,1)} \frac{xy}{y^2-1} = \lim_{x\to 0} \frac{mx^2+x}{m^2x^2+2mx}$$

$$= \frac{1}{2m}$$

Limit depends on m, so discontinuous at (0,3) by two path rule.

Chapter 2

2.1: $x + 2y$	**2.2**: $x^2 + 4xy + 3y^2$
$\frac{\partial f}{\partial x} = 1; \frac{\partial^2 f}{\partial x^2} = 0; \frac{\partial^2 f}{\partial y \partial x} = 0$	$\frac{\partial f}{\partial x} = 2x + 4y; \frac{\partial^2 f}{\partial x^2} = 2; \frac{\partial^2 f}{\partial y \partial x} = 4$
$\frac{\partial f}{\partial y} = 2; \frac{\partial^2 f}{\partial y^2} = 0; \frac{\partial^2 f}{\partial x \partial y} = 0$	$\frac{\partial f}{\partial y} = 4x + 6y; \frac{\partial^2 f}{\partial y^2} = 6; \frac{\partial^2 f}{\partial x \partial y} = 4$
2.3: $x^2 + yx$	**2.4**: $\sqrt{x^2 + y^2}$
$\frac{\partial f}{\partial x} = 2x + y; \frac{\partial^2 f}{\partial x^2} = 2; \frac{\partial^2 f}{\partial y \partial x} = 1$	$\frac{\partial f}{\partial x} = \frac{x}{\sqrt{x^2+y^2}}; \frac{\partial f}{\partial y} = \frac{y}{\sqrt{x^2+y^2}};$
$\frac{\partial f}{\partial y} = x; \frac{\partial^2 f}{\partial y^2} = 0; \frac{\partial^2 f}{\partial x \partial y} = 1$	$\frac{\partial^2 f}{\partial x^2} = \frac{\sqrt{x^2+y^2} - x^2/\sqrt{x^2+y^2}}{x^2+y^2};$
	$\frac{\partial^2 f}{\partial y \partial x} = \frac{\partial^2 f}{\partial x \partial y} = \frac{-xy}{\sqrt{x^2+y^2}^3};$
	$\frac{\partial^2 f}{\partial y^2} = \frac{\sqrt{x^2+y^2} - y^2/\sqrt{x^2+y^2}}{x^2+y^2}$
2.5: $\sin x + \cos y$	**2.6**: $y^2 \sin 3x + x^2 \cos 4y$
$\frac{\partial f}{\partial x} = \cos x;$	$\frac{\partial f}{\partial x} = 3y^2 \cos 3x + 2x \cos 4y;$
$\frac{\partial^2 f}{\partial x^2} = -\sin x; \frac{\partial^2 f}{\partial y \partial x} = 0$	$\frac{\partial^2 f}{\partial x^2} = -9y^2 \sin x + 2 \cos 4y;$
$\frac{\partial f}{\partial y} = -\sin y;$	$\frac{\partial^2 f}{\partial y \partial x} = 6y \cos x - 8x \sin 4y;$
$\frac{\partial^2 f}{\partial y^2} = -\cos y; \frac{\partial^2 f}{\partial x \partial y} = 0$	$\frac{\partial f}{\partial y} = 2y \sin 3x - 4x^2 \sin 4y;$
	$\frac{\partial^2 f}{\partial y^2} = 2 \sin 3x - 16x^2 \cos 4y;$
	$\frac{\partial^2 f}{\partial x \partial y} = 6y \cos 3x - 8x \sin 4y$

2.7: e^{xy}

$$\frac{\partial f}{\partial x} = ye^{xy};$$

$$\frac{\partial^2 f}{\partial x^2} = y^2 e^{xy};$$

$$\frac{\partial^2 f}{\partial y \partial x} = e^{xy} + xy\, e^{xy}$$

$$\frac{\partial f}{\partial y} = xe^{xy};$$

$$\frac{\partial^2 f}{\partial y^2} = x^2 e^{xy};$$

$$\frac{\partial^2 f}{\partial x \partial y} = e^{xy} + xy\, e^{xy}$$

2.8: $\ln(x^2 + y^2)$

$$\frac{\partial f}{\partial x} = \frac{2x}{x^2+y^2};$$

$$\frac{\partial^2 f}{\partial x^2} = \frac{2(x^2+y^2)-4x^2}{(x^2+y^2)^2};$$

$$\frac{\partial^2 f}{\partial y \partial x} = \frac{-4xy}{(x^2+y^2)^2};$$

$$\frac{\partial f}{\partial y} = \frac{2y}{x^2+y^2};$$

$$\frac{\partial^2 f}{\partial y^2} = \frac{2(x^2+y^2)-4y^2}{(x^2+y^2)^2};$$

$$\frac{\partial^2 f}{\partial x \partial y} = \frac{-4xy}{(x^2+y^2)^2}$$

2.9: $xy\cos 2x$

$$\frac{\partial f}{\partial x} = y\cos 2x - 2xy\sin 2x;$$

$$\frac{\partial^2 f}{\partial x^2} = -2y\sin 2x - 2y\sin 2x$$
$$-4xy\cos 2x;$$

$$\frac{\partial^2 f}{\partial y \partial x} = \cos 2x - 2x\sin 2x;$$

$$\frac{\partial f}{\partial y} = x\cos 2x;$$

$$\frac{\partial^2 f}{\partial y^2} = 0;$$

$$\frac{\partial^2 f}{\partial x \partial y} = \cos 2x - 2x\sin 2x$$

2.10: $\tan^{-1}(x^2 + y^2)$

$$\frac{\partial f}{\partial x} = \frac{2x}{1+x^2+y^2};$$

$$\frac{\partial^2 f}{\partial x^2} = \frac{2(1+x^2+y^2)-4x^2}{(1+x^2+y^2)^2};$$

$$\frac{\partial^2 f}{\partial y \partial x} = \frac{-4xy}{(1+x^2+y^2)^2};$$

$$\frac{\partial f}{\partial y} = \frac{2y}{1+x^2+y^2};$$

$$\frac{\partial^2 f}{\partial y^2} = \frac{2(1+x^2+y^2)-4y^2}{(1+x^2+y^2)^2};$$

$$\frac{\partial^2 f}{\partial x \partial y} = \frac{-4xy}{(1+x^2+y^2)^2}$$

2.11: $4x^3y + 3y^2$

$$\frac{\partial f}{\partial x} = 12x^2y; \quad \frac{\partial^2 f}{\partial x^2} = 24xy;$$

$$\frac{\partial^3 f}{\partial x^3} = 24y;$$

$$\frac{\partial f}{\partial y} = 4x^3 + 6y;$$

$$\frac{\partial^2 f}{\partial y^2} = 6; \quad \frac{\partial^3 f}{\partial y^3} = 0;$$

$$\frac{\partial^2 f}{\partial x \partial y} = 12x^2; \quad \frac{\partial^3 f}{\partial x^2 \partial y} = 24x;$$

2.12: $(x + 1)^3 yz + (2y - 1)^2 xz$

$$\frac{\partial f}{\partial x} = 3yz(x + 1)^2 + (2y - 1)^2 z;$$

$$\frac{\partial^2 f}{\partial x^2} = 6yz(x + 1);$$

$$\frac{\partial^3 f}{\partial x^3} = 6yz;$$

$$\frac{\partial f}{\partial y} = z(x + 1)^3 + 2xz(2y - 1);$$

$$\frac{\partial^2 f}{\partial y^2} = 4xz; \quad \frac{\partial^3 f}{\partial y^3} = 0;$$

$$\frac{\partial^2 f}{\partial x \partial y} = 3z(x + 1)^2 + 2z(2y - 1);$$

$$\frac{\partial^3 f}{\partial x^2 \partial y} = 6z(x + 1);$$

2.13: $\sin xy$

$$\frac{\partial f}{\partial x} = y \cos xy; \quad \frac{\partial^2 f}{\partial x^2} = -y^2 \sin xy;$$

$$\frac{\partial^3 f}{\partial x^3} = -y^3 \cos xy;$$

$$\frac{\partial f}{\partial y} = x \cos xy; \quad \frac{\partial^2 f}{\partial y^2} = -x^2 \sin xy;$$

$$\frac{\partial^3 f}{\partial y^3} = -x^3 \cos xy;$$

$$\frac{\partial^2 f}{\partial x \partial y} = \cos xy - xy \sin xy;$$

$$\frac{\partial^3 f}{\partial x^2 \partial y} = -y \sin xy - y \sin xy$$
$$-xy^2 \cos xy;$$

2.14: $e^{x^2 - y^2}$

$$\frac{\partial f}{\partial x} = 2xe^{x^2 - y^2};$$

$$\frac{\partial^2 f}{\partial x^2} = 2e^{x^2 - y^2} + 4x^2 e^{x^2 - y^2};$$

$$\frac{\partial^3 f}{\partial x^3} = \left[e^{x^2 - y^2}\right][4x + 8x + 8x^3];$$

$$\frac{\partial f}{\partial y} = -2ye^{x^2 - y^2};$$

$$\frac{\partial^2 f}{\partial y^2} = -2e^{x^2 - y^2} + 4y^2 e^{x^2 - y^2};$$

$$\frac{\partial^3 f}{\partial y^3} = \left[e^{x^2 - y^2}\right][4y + 8y - 8y^3];$$

$$\frac{\partial^2 f}{\partial x \partial y} = -4xye^{x^2 - y^2};$$

$$\frac{\partial^3 f}{\partial x^2 \partial y} = -16x^2 ye^{x^2 - y^2}$$
$$-4ye^{x^2 - y^2}$$

Chapter 3

3.1: $w = x^2y - y^2x$ $$dw = \frac{\partial w}{\partial x}dx + \frac{\partial w}{\partial y}dy$$ $$= (2xy - y^2)dx + (x^2 - 2xy)dy$$ $$= 1(0.01) + (-1)(0.01) = 0$$	**3.2**: $w = 4x^2 + 2xy + y^2$ $$dw = (8x + 2y)dx + (2x + 2y)dy$$ $$= (10)(0.01) + (4)(0.0.1) = 0.41$$
3.3: $w = \frac{xy}{x+y}$ $$= \left(\frac{xy+y^2-xy}{(x+y)^2}\right)dx + \left(\frac{xy+x^2-xy}{(x+y)^2}\right)dy$$ $$= 2(0.25)(0.01) = 0.005$$	**3.4**: $w = y\sin(1-x)$ $$dw = (-y\cos(1-x))dx + (\sin(1-x))dy$$ $$= (-1)(0.01) = -0.01$$
3.5: $w = e^{x^2y}$ $$dw = \left(2xye^{x^2y}\right)dx + \left(x^2e^{x^2y}\right)dy$$ $$= (2e)(0.01) + (e)(0.0.1) = 0.08$$	**3.6**: $w = \ln(2x - y)$ $$dw = \left(\frac{2}{2x-y}\right)dx + \left(\frac{-1}{2x-y}\right)dy$$ $$= (2)(0.01) - (0.01) = 0.01$$
3.7: $u = xy;\ v = x^2y^2;$ $$w = 2u - 3v$$ $$\frac{\partial u}{\partial x} = y;\ \frac{\partial w}{\partial u} = 2;$$ $$\frac{\partial v}{\partial x} = 2xy^2;\ \frac{\partial w}{\partial v} = -3;$$ $$\frac{\partial w}{\partial x} = \frac{\partial w}{\partial u}\frac{\partial u}{\partial x} + \frac{\partial w}{\partial v}\frac{\partial v}{\partial x}$$ $$= 2y - 6xy^2$$ $$w = 2xy - 3x^2y^2;$$ $$\frac{\partial w}{\partial x} = 2y - 6xy^2$$	**3.8**: $u = x - y;\ v = \ln x;$ $$w = uv - u^2$$ $$\frac{\partial u}{\partial x} = 1;\ \frac{\partial w}{\partial u} = v - 2u;$$ $$\frac{\partial v}{\partial x} = 1/x;\ \frac{\partial w}{\partial v} = u;$$ $$\frac{\partial w}{\partial x} = v - 2u + u/x$$ $$= \ln x - 2x - 2y + \frac{x-y}{x}$$ $$w = (x - y)\ln x - (x - y)^2;$$ $$\frac{\partial w}{\partial x} = \ln x + \frac{x-y}{x} - 2(x - y)$$

3.9: $u = x^2$; $v = 2x + y$;
$$w = u/v$$

$$\frac{\partial u}{\partial x} = 2x; \ \frac{\partial w}{\partial u} = 1/v;$$

$$\frac{\partial v}{\partial x} = 2; \ \frac{\partial w}{\partial v} = -u/v^2;$$

$$\frac{\partial w}{\partial x} = \frac{2x}{v} - \frac{2u}{v^2} = \frac{2x}{2x+y} - \frac{2x^2}{(2x+y)^2}$$

$$= \frac{4x^2 + 2xy - 2x^2}{(2x+y)^2}$$

$$w = \frac{x^2}{2x+y}; \ \frac{\partial w}{\partial x} = \frac{(2x+y)2x - 2x^2}{(2x+y)^2}$$

3.10: $u = \cos 2x$; $v = \sin y$;
$$w = uv$$

$$\frac{\partial u}{\partial x} = -2\sin 2x; \ \frac{\partial w}{\partial u} = v;$$

$$\frac{\partial v}{\partial x} = 0; \ \frac{\partial w}{\partial v} = u;$$

$$\frac{\partial w}{\partial x} = -2\sin y \sin 2x$$

$$w = \cos 2x \sin y;$$

$$\frac{\partial w}{\partial x} = -2\sin 2x \sin y$$

3.11: $u = e^{xy}$; $v = x + y$;
$$w = 4u^2 + v$$

$$\frac{\partial u}{\partial x} = ye^{xy}; \ \frac{\partial w}{\partial u} = 8u;$$

$$\frac{\partial v}{\partial x} = 1; \ \frac{\partial w}{\partial v} = 1;$$

$$\frac{\partial w}{\partial x} = 8ye^{2xy} + 1$$

$$w = 4e^{2xy} + x + y;$$

$$\frac{\partial w}{\partial x} = 8ye^{2xy} + 1$$

3.12: $u = xy$; $v = x - y$;
$$w = \tan^{-1}(2u + v)$$

$$\frac{\partial u}{\partial x} = y; \ \frac{\partial w}{\partial u} = \frac{2}{1+(2u+v)^2};$$

$$\frac{\partial v}{\partial x} = 1; \ \frac{\partial w}{\partial v} = \frac{1}{1+(2u+v)^2};$$

$$\frac{\partial w}{\partial x} = \frac{2y+1}{1+(2xy+x-y)^2}$$

$$w = \tan^{-1}(2xy + x - y);$$

$$\frac{\partial w}{\partial x} = \frac{2y+1}{1+(2xy+x-y)^2}$$

3.13: $x^4 y = y^3 x$

$$x^4 y - y^3 x = 0;$$

$$y' = \frac{-\partial f/\partial x}{\partial f/\partial y} = \frac{-4yx^3 + y^3}{x^4 - 3xy^2}$$

3.14: $x^2 y^2 = xy + \cos x$

$$x^2 y^2 - xy - \cos x = 0;$$

$$y' = \frac{-2xy^2 + y - \sin x}{2yx^2 - x}$$

Chapter 4

4.1: $x^2 - 2xy + 4y^2$ $f_x = 2x - 2y = 0 \rightarrow x = y;$ $f_y = -2x + 8y = 0 \rightarrow x = 4y;$ Critical point is $(0,0)$. $f_{xx} = 2 > 0;\ f_{yy} = 8;\ f_{xy} = -2;$ $f_{xx}f_{yy} - f_{xy}^{\,2} = 12 > 0;$ Point is a minimum.	**4.2:** $x^2y - y^2x$ $f_x = 2xy - y^2 = 0 \rightarrow x = y/2;$ $f_y = -2xy + x^2 = 0 \rightarrow x = 2y;$ Critical point is $(0,0)$. $f_{xx} = 2y = 0;\ f_{yy} = -2x = 0;$ $f_{xy} = 2x - 2y = 0;$ $f_{xx}f_{yy} - f_{xy}^{\,2} = 0;$ Point is an inflection.
4.3: $(x-1)^2 + (y-2)^2$ $f_x = 2x - 2 = 0 \rightarrow x = 1;$ $f_y = 2y - 4 = 0 \rightarrow y = 2;$ Critical point is $(1,2)$. $f_{xx} = 2 > 0;\ f_{yy} = 2;\ f_{xy} = 0;$ $f_{xx}f_{yy} - f_{xy}^{\,2} = 4 > 0;$ Point is a minimum.	**4.4:** $y^3 - x^2y$ $f_x = -2xy = 0 \rightarrow x\ or\ y = 0;$ $f_y = 3y^2 - x^2 = 0 \rightarrow y = 0$ if $x = 0;$ Critical point is $(0,0)$. $f_{xx} = -2y > 0;\ f_{yy} = 6y = 0;$ $f_{xy} = -2x = 0;$ $f_{xx}f_{yy} - f_{xy}^{\,2} = 0;$ Point is an inflection.
4.5: $(4x+2)^2 + y^2$ $f_x = 8x + 4 = 0 \rightarrow x = -1/2;$ $f_y = 2y = 0 \rightarrow y = 0;$ Critical point is $(-\frac{1}{2}, 0)$. $f_{xx} = 8 > 0;\ f_{yy} = 2;\ f_{xy} = 0;$ $f_{xx}f_{yy} - f_{xy}^{\,2} = 16 > 0;$ Point is a minimum.	**4.6:** $e^{x^2+y^2}$ $f_x = 2xe^{x^2+y^2} = 0 \rightarrow x = 0;$ $f_y = 2ye^{x^2+y^2} = 0 \rightarrow y = 0;$ Critical point is $(0,0)$. $f_{xx} = (2 + 4x^2)e^{x^2+y^2} = 2 > 0;$ $f_{yy} = (2 + 4y^2)e^{x^2+y^2} = 2;$ $f_{xy} = 4xye^{x^2+y^2} = 0;$ $f_{xx}f_{yy} - f_{xy}^{\,2} = 4 > 0;$ Point is a minimum.

4.7: $\cos^2 x + \cos^2 y$	**4.8:** $x^2 y^2 - 2xy + (y-1)^2$
$f_x = -2\cos x \sin x = 0 \rightarrow x = 0;$ $f_y = -2\cos y \sin y = 0 \rightarrow y = 0;$ Critical point is $(0,0)$. $f_{xx} = -2(\cos^2 x - \sin^2 x) = -2;$ $f_{yy} = -2(\cos^2 x - \sin^2 x) = -2;$ $f_{xy} = 0;$ $f_{xx}f_{yy} - f_{xy}^2 = 4 > 0;$ Point is a maximum $(f_{xx} < 0)$.	$f_x = 2xy^2 - 2y = 0 \rightarrow x = 1/y;$ $f_y = 2yx^2 - 2x + 2y - 2 = 0 \rightarrow$ $y = 1 \rightarrow x = 1;$ Critical point is $(1,1)$. $f_{xx} = 2y^2 = 2 > 0;$ $f_{yy} = 2x^2 + 2 = 4;$ $f_{xy} = 4xy - 2 = 2;$ $f_{xx}f_{yy} - f_{xy}^2 = 4 > 0;$ Point is a minimum.
4.9: $x^3 + 3xy - y^3$	**4.10:** $x^3 y - 3xy - y^2$
$f_x = 3x^2 + 3y = 0;$ $f_y = 3x - 3y^2 = 0;$ Critical points: $(1,-1)$ and $(0,0)$. At $(1,-1)$: $f_{xx} = 6x = 6 > 0;$ $f_{yy} = -6y = 6; \ f_{xy} = 3;$ $f_{xx}f_{yy} - f_{xy}^2 = 27 > 0;$ At $(0,0)$: $f_{xx} = 0; \ f_{yy} = 0; \ f_{xy} = 3;$ $f_{xx}f_{yy} - f_{xy}^2 = -9 < 0;$ Points are a minimum for $(1,-1)$ and an infection for $(0,0)$.	$f_x = 3x^2 y - 3y = 0;$ $f_y = x^3 - 3x - 2y = 0;$ Critical points: $(1,-1), (-1,1)$ and $(0,0)$. At $(1,-1)$ and $(-1,1)$: $f_{xx} = 6xy = -6 < 0;$ $f_{yy} = -2; \ f_{xy} = 3x^2 - 3 = 0;$ $f_{xx}f_{yy} - f_{xy}^2 = 12 > 0;$ At $(0,0)$: $f_{xx} = 0; \ f_{yy} = -2; \ f_{xy} = -3;$ $f_{xx}f_{yy} - f_{xy}^2 = -9 < 0;$ Points are maxima for $(1,-1)$ and $(-1,1)$; an infection for $(0,0)$.

4.11: $x^3 - 3x^2y + y^4$

$$f_x = 3x^2 - 6xy = 0;$$
$$f_y = -3x^2 + 4y^3 = 0;$$

Critical points: $(6,3)$ and $(0,0)$.

At $(6,3)$:
$$f_{xx} = 6x - 6y = 18 > 0;$$
$$f_{yy} = 12y^2 = 108;$$
$$f_{xy} = -6x = -36;$$
$$f_{xx}f_{yy} - f_{xy}^2 = 648 > 0;$$

At $(0,0)$:
$$f_{xx} = 0; \quad f_{yy} = 0; \quad f_{xy} = 0;$$
$$f_{xx}f_{yy} - f_{xy}^2 = 0;$$

Points are a minimum for $(6,3)$ and an infection for $(0,0)$.

4.12: $x^3 - 3xy + xy^3$

$$f_x = 3x^2 - 3y + y^3 = 0;$$
$$f_y = -3x + 3xy^2 = 0;$$

Critical points: $\left(\pm\sqrt{2/3}, 1\right)$ and $(0,0)$.

At $\left(\pm\sqrt{2/3}, 1\right)$:
$$f_{xx} = 6x = \pm6\sqrt{2/3};$$
$$f_{yy} = 6xy = \pm6\sqrt{2/3};$$
$$f_{xy} = -3 + 3y^2 = 0;$$
$$f_{xx}f_{yy} - f_{xy}^2 = 24 > 0;$$

At $(0,0)$:
$$f_{xx} = 0; \quad f_{yy} = 0; \quad f_{xy} = -3;$$
$$f_{xx}f_{yy} - f_{xy}^2 = -9 < 0;$$

Minimum for $\left(\sqrt{2/3}, 1\right)$, maximum for $\left(-\sqrt{2/3}, 1\right)$, niether for $(0,0)$.

4.13: $x^2y^2 - 2xy$

$$f_x = 2xy^2 - 2y = 0;$$
$$f_y = 2yx^2 - 2x = 0;$$

Critical points: $(1,1)$, $(-1,-1)$ and $(0,0)$.

At $(1,1)$ and $(-1,-1)$:
$$f_{xx} = 2y^2 = 2 > 0;$$
$$f_{yy} = 2x^2 = 2;$$
$$f_{xy} = 4xy - 2 = 2;$$
$$f_{xx}f_{yy} - f_{xy}^2 = 0;$$

At $(0,0)$: $f_{xx}f_{yy} - f_{xy}^2 = -4 < 0$;

All critical points are inflections.

4.14: $x^4y^2 - x^2 - y^2$

$$f_x = 4x^3y^2 - 2x = 0;$$
$$f_y = 2yx^4 - 2y = 0;$$

Critical points: $\left(\pm1, \pm1/\sqrt{2}\right)$ (four points), and $(0,0)$.

At $\left(\pm1, \pm1/\sqrt{2}\right)$
$$f_{xx} = 12x^2y^2 - 2 = 4 > 0;$$
$$f_{yy} = 2x^4 - 2 = 0;$$
$$f_{xy} = 8x^3y = \pm8/\sqrt{2};$$
$$f_{xx}f_{yy} - f_{xy}^2 = -32 < 0;$$

At $(0,0)$: $f_{xx} = -2 < 0$
$$f_{xx}f_{yy} - f_{xy}^2 = 4 > 0;$$

All inflections except a max at $(0,0)$.

Chapter 5

5.1: $\int_0^3 \int_0^2 x^2 + y^2 \, dx \, dy$

$\int_0^3 \frac{x^3}{3} + xy^2 \Big|_0^2 \, dy = \int_0^3 \frac{8}{3} + 2y^2 \, dy$

$= \frac{8y}{3} + \frac{2y^3}{3} \Big|_0^3 = 8 + 18 = 26.$

$\int_0^2 yx^2 + \frac{y^3}{3} \Big|_0^3 \, dx = \int_0^2 3x^2 + 9 \, dx$

$= x^3 + 9x \Big|_0^2 = 8 + 18 = 26.$

5.2: $\int_1^2 \int_0^2 4xy + 3x^2 \, dx \, dy$

$\int_1^2 2yx^2 + x^3 \Big|_0^2 \, dy = \int_1^2 8y + 8 \, dy =$

$4y^2 + 8y \Big|_1^2 = 16 + 16 - 4 - 8 = 20.$

$\int_0^2 2xy^2 + 3yx^2 \Big|_1^2 \, dx$

$= \int_0^2 8x + 6x^2 - 2x - 3x^2 \, dx$

$= 3x^2 + x^3 \Big|_0^2 = 12 + 8 = 20.$

5.3: $\int_0^1 \int_0^1 3x^2y + 2(x-1)y^2 \, dx dy$

$\int_0^1 yx^3 + y^2(x-1)^2 \Big|_0^1 \, dy$

$= \int_0^1 y - y^2 \, dy = \frac{y^2}{2} - \frac{y^3}{3} \Big|_0^1 = \frac{1}{6}.$

$\int_0^1 \frac{3y^2x^2}{2} + \frac{2y^3(x-1)}{3} \Big|_0^1 \, dx$

$= \int_0^1 \frac{3x^2}{2} + \frac{2(x-1)}{3} \, dx$

$= \frac{x^3}{2} + \frac{(x-1)^2}{3} \Big|_0^1 = \frac{1}{2} - \frac{1}{3} = \frac{1}{6}.$

5.4: $\int_0^{\pi/2} \int_0^\pi \sin x \cos y \, dx \, dy$

$\int_0^{\pi/2} - \cos x \cos y \Big|_0^\pi \, dy$

$= \int_0^{\pi/2} 2 \cos y \, dy = 2 \sin y \Big|_0^{\pi/2} = 2.$

$\int_0^\pi \sin x \sin y \Big|_0^{\pi/2} \, dx = \int_0^\pi \sin x \, dx$

$= - \cos x \Big|_0^\pi = -(-1) - (-1) = 2.$

5.5: $\int_0^3 \int_0^1 y^2 e^{2x} \, dx \, dy$

$\int_0^3 \frac{y^2 e^{2x}}{2} \Big|_0^1 \, dy = \int_0^3 \frac{y^2(e-1)}{2} \, dy$

$= \frac{y^3(e-1)}{6} \Big|_0^3 = \frac{9(e-1)}{2}.$

$\int_0^1 \frac{y^3 e^{2x}}{3} \Big|_0^3 \, dx = \int_0^1 9e^{2x} \, dx = \frac{9(e-1)}{2}.$

5.6: $\int_0^1 \int_0^\pi y \sin x \, dx \, dy$

$\int_0^1 -y \cos x \Big|_0^\pi \, dy$

$= \int_0^1 2y \, dy = y^2 \Big|_0^1 = 1.$

$\int_0^\pi \frac{y^2 \sin x}{2} \Big|_0^1 \, dx = - \cos x \Big|_0^\pi = 1.$

5.7: $\int_0^1 \int_1^2 (2x-4)^2(4y+2)\, dx\, dy$

$$\int_0^1 \frac{(2x-4)^3(4y+2)}{6}\Big|_1^2\, dy$$

$$= \int_0^1 \frac{8(4y+2)}{6}\Big|_1^2\, dy$$

$$= \frac{(4y+2)^2}{3}\Big|_0^1 = 12 - \frac{4}{3} = 10.7.$$

$$\int_1^2 \frac{(4y+2)^2(2x-4)^2}{8}\Big|_0^1\, dx$$

$$= \int_1^2 \frac{32(2x-4)^2}{8}\, dx$$

$$= \frac{4(2x-4)^3}{3}\Big|_1^2 = \frac{32}{3} = 10.7.$$

5.8: $\int_0^1 \int_0^1 x(xy-1)^2\, dx\, dy$

Let $u = (xy-1)^2$; $dv = x$;

$du = 2xy^2 - 2y$; $v = \frac{x^2}{2}$;

$\int u\, dv = uv - \int v\, du \rightarrow$

$$\int_0^1 \left[\frac{x^2(xy-1)^2}{2} - \int_0^1 x^3y^2 - x^2y\, dx\right]\Big|_0^1\, dy$$

$$= \int_0^1 \left[\frac{x^2(xy-1)^2}{2} - \frac{x^4y^2}{4} + \frac{x^3y}{3}\right]\Big|_0^1\, dy$$

$$= \int_0^1 \left[\frac{(y-1)^2}{2} - \frac{y^2}{4} + \frac{y}{3}\right]\, dy$$

$$= \frac{(y-1)^3}{6} - \frac{y^3}{12} + \frac{y^2}{6}\Big|_0^1 = \frac{1}{12} + \frac{1}{6} = \frac{1}{4}.$$

$$\int_0^1 \frac{(xy-1)^3}{3}\Big|_0^1\, dx = \int_0^1 \frac{(x-1)^3}{3} + \frac{1}{3}\, dx$$

$$= \frac{(x-1)^4}{12} + \frac{x}{3}\Big|_0^1 = \frac{1}{3} - \frac{1}{12} = \frac{1}{4}.$$

5.9: $f(x,y) = x^3 + yx$;

$y = x$; $y = x^2$; $0 \le x \le 1$

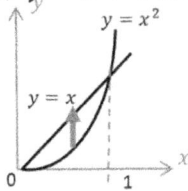

$$\int_0^1 \int_{x^2}^x x^3 + yx\, dy\, dx$$

$$\int_0^1 x^3y + \frac{xy^2}{2}\Big|_{x^2}^x\, dx$$

$$= \int_0^1 x^4 + \frac{x^3}{2} - x^5 - \frac{x^5}{2}\, dx$$

$$= \frac{x^5}{5} + \frac{x^4}{8} - \frac{x^6}{6} - \frac{x^6}{12}\Big|_0^1 = 0.075.$$

5.10: Reverse the order of integration for Problem 5.9

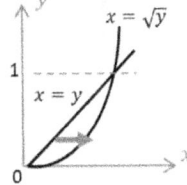

$$\int_0^1 \int_y^{\sqrt{y}} x^3 + yx\, dx\, dy$$

$$\int_0^1 \frac{x^4}{4} + \frac{yx^2}{2}\Big|_y^{\sqrt{y}}\, dy$$

$$= \int_0^1 \frac{y^2}{4} + \frac{y^2}{2} - \frac{y^4}{4} - \frac{y^3}{2}\, dy$$

$$= \frac{y^3}{12} + \frac{y^3}{6} - \frac{y^5}{20} - \frac{y^4}{8}\Big|_0^1 = 0.075.$$

5.11: $f(x, y) = xy$;
$\quad x = 0; \quad x = 2y; \quad 0 \leq y \leq 1$

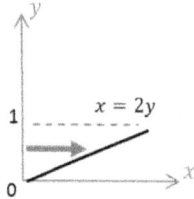

$$\int_0^1 \int_0^{2y} xy \, dx \, dy$$

$$\int_0^1 \frac{yx^2}{2} \Big|_0^{2y} dy = \int_0^1 2y^3 \, dy$$

$$= \frac{y^4}{2} \Big|_0^1 = 0.5.$$

5.12: Reverse the order of integration for Problem 5.11

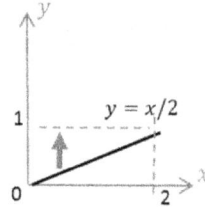

$$\int_0^2 \int_{x/2}^1 xy \, dy \, dx$$

$$\int_0^2 \frac{xy^2}{2} \Big|_{x/2}^1 dx = \int_0^2 \frac{x}{2} - \frac{x^3}{8} \, dx$$

$$= \frac{x^2}{4} - \frac{x^4}{32} \Big|_0^2 = 1 - \frac{1}{2} = 0.5.$$

5.13: $f(x, y) = x^2 y$;
$\quad y = x; \quad y = 2x; \quad 0 \leq x \leq 2$

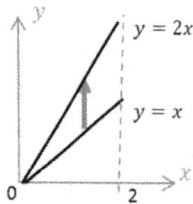

$$\int_0^2 \int_x^{2x} x^2 y \, dy \, dx$$

$$\int_0^2 \frac{x^2 y^2}{2} \Big|_x^{2x} dx$$

$$= \int_0^2 2x^4 - \frac{x^4}{2} \, dx$$

$$= \frac{2x^5}{5} - \frac{x^5}{10} \Big|_0^2 = 9.6.$$

5.14: Reverse the order of integration for Problem 5.13

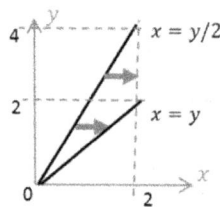

$$\int_0^2 \int_{y/2}^y x^2 y \, dx \, dy + \int_2^4 \int_{y/2}^2 x^2 y \, dx \, dy$$

$$\int_0^2 \frac{x^3 y}{3} \Big|_{y/2}^y dy + \int_2^4 \frac{x^3 y}{3} \Big|_{y/2}^2 dy$$

$$= \int_0^2 \frac{y^4}{3} - \frac{y^4}{24} \, dy + \int_2^4 \frac{8y}{3} - \frac{y^4}{24} \, dy$$

$$= \frac{y^5}{15} - \frac{y^5}{120} \Big|_0^2 + \frac{4y^2}{3} - \frac{y^5}{120} \Big|_2^4$$

$$= \frac{32}{15} - \frac{32}{120} + \frac{64}{3} - \frac{1024}{120} - \frac{16}{3} + \frac{32}{120}$$

$$= 9.6.$$

Chapter 6

6.1: $(4, 5, -2)$ $$r = \sqrt{4^2 + 5^2} = \sqrt{41} = 6.4;$$ $$\theta = \tan^{-1}\frac{5}{4} = 0.9; \ z = -2$$ $$(6.4, 0.9, -2)$$	**6.2**: $(7, 2, 12)$ $$r = \sqrt{7^2 + 2^2} = \sqrt{53} = 7.3;$$ $$\theta = \tan^{-1}\frac{7}{2} = 1.3; \ z = 12$$ $$(7.3, 1.3, 12)$$
6.3: $f(x, y, z) = 4x + y^2 z$ $$f(r, \theta, z) = 4r\cos\theta - z(r\sin\theta)^2$$	**6.4**: $f(x, y, z) = z(x^2 + y^2)$ $$f(r, \theta, z) = z[(r\cos\theta)^2 + (r\sin\theta)^2] = zr^2$$
6.5: $(4r^2 - 1)^2 \cos\theta$ $$\frac{\partial f}{\partial r} = 2(8r)\cos\theta\,(4r^2 - 1);$$ $$\frac{1}{r}\frac{\partial f}{\partial \theta} = \frac{1}{r}[-(4r^2 - 1)^2 \sin\theta]$$	**6.6**: $\dfrac{r^3}{\sin\theta}$ $$\frac{\partial f}{\partial r} = \frac{3r^2}{\sin\theta};$$ $$\frac{1}{r}\frac{\partial f}{\partial \theta} = -\frac{r^2 \cos\theta}{\sin^2\theta}$$
6.7: $\ln(r^2\theta)$ $$\frac{\partial f}{\partial r} = \frac{2r\theta}{r^2\theta} = \frac{2}{r};$$ $$\frac{1}{r}\frac{\partial f}{\partial \theta} = \frac{1}{r}\frac{r^2}{r^2\theta} = \frac{1}{r\theta}$$	**6.8**: $e^{2r\cos\theta}$ $$\frac{\partial f}{\partial r} = 2\cos\theta\, e^{2r\cos\theta};$$ $$\frac{1}{r}\frac{\partial f}{\partial \theta} = -2\sin\theta\, e^{2r\cos\theta}$$
6.9: $f(r, \theta) = r\theta - 1;$ $\quad 0 \le r \le 2; \ 0 \le \theta \le \pi$ $$\int_0^\pi \int_0^2 (r\theta - 1)\, r\,dr\,d\theta$$ $$= \int_0^\pi \frac{r^3\theta}{3} - \frac{r^2}{2}\Big\|_0^2 \, d\theta = \int_0^\pi \frac{8\theta}{3} - 2 \ d\theta$$ $$= \frac{4\theta^2}{3} - 2\theta\Big\|_0^\pi = \frac{4\pi^2}{3} - 2\pi$$	**6.10**: $f(r, \theta) = r^2\theta - 3\theta^2 r;$ $\quad 0 \le r \le 1; \ 0 \le \theta \le \pi$ $$\int_0^\pi \int_0^1 (r^2\theta - 3\theta^2 r)\, r\,dr\,d\theta$$ $$= \int_0^\pi \frac{r^4\theta}{4} - r^3\theta^2 \Big\|_0^1 \, d\theta$$ $$= \int_0^\pi \frac{\theta}{4} - \theta^2 \ d\theta = \frac{\theta^2}{8} - \frac{\theta^3}{3}\Big\|_0^\pi$$ $$= \frac{\pi^2}{8} - \frac{\pi^3}{3}$$

6.11: $f(r, \theta) = \cos r\theta$;
   $0 \le r \le 1/2; \ 0 \le \theta \le \pi/2$

This time we'll integrate ɤ θ first for convenience:

$$\int_0^{1/2} \int_0^{\pi/2} (\cos r\theta) \, r \, d\theta dr$$

$$u = r\theta; \ du = r$$

$$= \int_0^{1/2} \sin r\theta \, \big|_0^{\pi/2} \, dr$$

$$= \int_0^{1/2} \sin \frac{\pi r}{2} \, dr = \frac{-2}{\pi} \cos \frac{\pi r}{2} \Big|_0^{1/2}$$

$$= \frac{-2}{\pi} \left[\frac{\sqrt{2}}{2} - 1 \right] = \frac{2 - \sqrt{2}}{\pi}$$

6.12: $f(r, \theta) = \sin r^2 \cos \theta$;
   $0 \le r \le \sqrt{\pi/2}; \ 0 \le \theta \le \pi/2$

$$\int_0^{\pi/2} \int_0^{\sqrt{\pi/2}} (\sin r^2 \cos \theta) \, r \, dr d\theta$$

$$u = r^2; \ du = 2r$$

$$= \int_0^{\pi/2} \frac{-\cos r^2 \cos \theta}{2} \, \Big|_0^{\sqrt{\pi/2}} \, d\theta$$

$$= \int_0^{\pi/2} \frac{\cos \theta}{2} \, d\theta = \frac{\sin \theta}{2} \Big|_0^{\pi/2} = \frac{1}{2}$$

6.13: $f(r, \theta) = 4(r\theta - 1)^3$;
   $0 \le r \le 2; \ 0 \le \theta \le \pi$

$$\int_0^2 \int_0^{\pi} (4(r\theta - 1)^3) \, r \, d\theta dr$$

$$u = r\theta - 1; \ du = r$$

$$= \int_0^2 (r\theta - 1)^4 \, \big|_0^{\pi} \, dr$$

$$= \int_0^2 (r\pi - 1)^4 - 1 \, dr$$

$$= \frac{(r\pi - 1)^5}{5\pi} - r \Big|_0^2 = \frac{(2\pi - 1)^5 + 1}{5\pi} - 2$$

6.14: $f(r, \theta) = \theta e^{r^2}$;
   $0 \le r \le 1; \ 0 \le \theta \le \pi$

$$\int_0^{\pi} \int_0^1 (\theta e^{r^2}) \, r \, dr d\theta$$

$$u = r^2; \ du = 2r$$

$$= \int_0^{\pi} \frac{\theta e^{r^2}}{2} \, \big|_0^1 \, d\theta$$

$$= \int_0^{\pi} \frac{\theta(e - 1)}{2} \, d\theta = \frac{\theta^2 (e - 1)}{4} \Big|_0^{\pi}$$

$$= \frac{\pi^2 (e - 1)}{4}$$

Chapter 7

7.1: $(4, 5, -2)$	**7.2:** $(7, 2, 12)$
$\rho = \sqrt{4^2 + 5^2 + 2^2} = \sqrt{45} = 6.7;$	$\rho = \sqrt{7^2 + 2^2 + 12^2} = 14.0;$
$\varphi = \tan^{-1}\left(\dfrac{\sqrt{4^2 + 5^2}}{-2}\right) = -1.3;$	$\varphi = \tan^{-1}\left(\dfrac{\sqrt{7^2 + 2^2}}{12}\right) = 0.55;$
$\theta = \tan^{-1}\dfrac{5}{4} = 0.9;$	$\theta = \tan^{-1}\dfrac{7}{2} = 1.3;$
$(6.7, -1.3, 0.9)$	$(14.0, 0.55, 1.3)$

7.3: $f(x, y, z) = 4x + y^2 z$	**7.4:** $f(x, y, z) = z(x^2 + y^2)$
$f(\rho, \varphi, \theta) = 4\rho \sin\varphi \cos\theta$ $+ \rho^2 \sin^2\varphi \sin^2\theta (\rho \cos\varphi)$	$f(\rho, \varphi, \theta) =$ $\rho \cos\varphi \, (\rho^2 \cos^2\varphi \sin^2\theta$ $+ \rho^2 \sin^2\varphi \sin^2\theta)$
	$= \rho^3 \cos\varphi \, \sin^2\theta$

7.5: $(4\rho^2 - 1)^2 \cos\theta \sin\varphi$	**7.6:** $\dfrac{\rho^3 \sin\varphi}{\sin\theta}$
$\dfrac{\partial f}{\partial \rho} = 2(8\rho) \cos\theta \sin\varphi \, (4\rho^2 - 1);$	$\dfrac{\partial f}{\partial \rho} = \dfrac{3\rho^2 \sin\varphi}{\sin\theta};$
$\dfrac{1}{\rho}\dfrac{\partial f}{\partial \varphi} = \dfrac{\cos\theta}{\rho}[(4\rho^2 - 1)^2 \cos\varphi];$	$\dfrac{1}{\rho}\dfrac{\partial f}{\partial \varphi} = \dfrac{\rho^3 \cos\varphi}{\rho \sin\theta} = \dfrac{\rho^2 \cos\varphi}{\sin\theta};$
$\dfrac{1}{\rho \sin\varphi}\dfrac{\partial f}{\partial \theta} = \dfrac{1}{\rho}[-(4\rho^2 - 1)^2 \sin\theta]$	$\dfrac{1}{\rho \sin\varphi}\dfrac{\partial f}{\partial \theta} = -\dfrac{\rho^2 \cos\theta}{\sin^2\theta}$

7.7: $\ln(\rho^2 \theta \varphi)$	**7.8:** $e^{2\rho \sin\varphi \cos\theta}$
$\dfrac{\partial f}{\partial \rho} = \dfrac{2\rho\theta\varphi}{\rho^2\theta\varphi} = \dfrac{2}{\rho};$	$\dfrac{\partial f}{\partial \rho} = 2\sin\varphi \cos\theta \, e^{2\rho \sin\varphi \cos\theta};$
$\dfrac{1}{\rho}\dfrac{\partial f}{\partial \varphi} = \dfrac{1}{\rho}\dfrac{\rho^2\theta}{\rho^2\theta\varphi} = \dfrac{1}{\rho\varphi};$	$\dfrac{1}{\rho}\dfrac{\partial f}{\partial \varphi} = 2\cos\varphi \cos\theta \, e^{2\rho \sin\varphi \cos\theta};$
$\dfrac{1}{\rho \sin\varphi}\dfrac{\partial f}{\partial \theta} = \dfrac{1}{\rho \sin\varphi}\dfrac{\rho^2\varphi}{\rho^2\theta\varphi} = \dfrac{1}{\rho\theta \sin\varphi}$	$\dfrac{1}{\rho \sin\varphi}\dfrac{\partial f}{\partial \theta} = -2\sin\theta \, e^{2\rho \sin\varphi \cos\theta}$

7.9: $f(\rho,\varphi,\theta) = \rho\cos\varphi - 1;$
$\quad 0 \leq \rho \leq 2; \ 0 \leq \varphi \leq \pi/2$

$\int_0^{\pi/2} \int_0^2 (\rho\cos\varphi - 1)\, \rho d\rho d\varphi$

$= \int_0^{\pi/2} \left[\frac{\rho^3 \cos\varphi}{3} - \frac{\rho^2}{2}\right]\Big|_0^2 \, d\varphi$

$= \int_0^{\pi/2} \frac{8}{3}\cos\varphi - 2 \, d\varphi$

$= \frac{8\sin\varphi}{3} - 2\varphi\Big|_0^{\pi/2} = \frac{8}{3} - \pi$

7.10: $f(\rho,\varphi,\theta) = \rho^2\theta - 3\theta^2\rho;$
$\quad 0 \leq \rho \leq 1; \ 0 \leq \theta \leq \pi$

$\int_0^{\pi} \int_0^1 (\rho^2\theta - 3\theta^2\rho)\, \rho\sin\varphi\, d\rho d\theta$

$= \int_0^{\pi} \left[\frac{\rho^4\theta}{4} - r^3\theta^2\right]\sin\varphi\,\Big|_0^1 \, d\theta$

$= \int_0^{\pi} \left[\frac{\theta}{4} - \theta^2\right]\sin\varphi \, d\theta$

$= \left[\frac{\theta^2}{8} - \frac{\theta^3}{3}\right]\sin\varphi\,\Big|_0^{\pi} = \left[\frac{\pi^2}{8} - \frac{\pi^3}{3}\right]\sin\varphi$

7.11: $f(\rho,\varphi,\theta) = \cos\rho\varphi\, ;$
$\quad 0 \leq \rho \leq 1/2; \ 0 \leq \varphi \leq \pi/2$

$\int_0^{1/2} \int_0^{\pi/2} (\cos\rho\varphi)\, \rho d\varphi d\rho$

$u = \rho\varphi; \ du = \rho$

$= \int_0^{1/2} \sin\rho\varphi\,\Big|_0^{\pi/2}\, d\rho = \int_0^{1/2} \sin\frac{\pi\rho}{2}\, d\rho$

$= \frac{-2}{\pi}\cos\frac{\pi\rho}{2}\Big|_0^{1/2} = \frac{-2}{\pi}\left[\frac{\sqrt{2}}{2} - 1\right] = \frac{2-\sqrt{2}}{\pi}$

7.12: $f(\rho,\varphi,\theta) = \sin\rho^2\cos\theta\, ;$
$\quad 0 \leq \rho \leq \sqrt{\pi/2}; \ 0 \leq \theta \leq \pi/2$

$\int_0^{\pi/2} \int_0^{\sqrt{\pi/2}} (\sin\rho^2\cos\theta)\, \rho\sin\varphi\, d\rho d\theta$

$u = \rho^2; \ du = 2\rho$

$= \int_0^{\pi/2} \frac{-\cos\rho^2\cos\theta\sin\varphi}{2}\,\Big|_0^{\sqrt{\pi/2}} \, d\theta$

$= \int_0^{\pi/2} \frac{\cos\theta\sin\varphi}{2}\, d\theta = \frac{\sin\theta\sin\varphi}{2}\,\Big|_0^{\pi/2} = \frac{1}{2}$

7.13: $f(\rho,\varphi,\theta) = 4(\rho\varphi - 1)^3;$
$\quad 0 \leq \rho \leq 2; \ 0 \leq \varphi \leq \pi$

$\int_0^2 \int_0^{\pi} (4(\rho\varphi - 1)^3)\, \rho d\varphi d\rho$

$u = \rho\varphi - 1; \ du = \rho$

$= \int_0^2 (\rho\varphi - 1)^4\,\Big|_0^{\pi} \, d\rho$

$= \int_0^2 (\rho\pi - 1)^4 - 1 \, d\rho$

$= \frac{(\rho\pi-1)^5}{5\pi} - \rho\Big|_0^2 = \frac{(2\pi-1)^5 + 1}{5\pi} - 2$

7.14: $f(\rho,\varphi,\theta) = \theta e^{\rho^2\sin\varphi};$
$\quad 0 \leq \rho \leq 1; \ 0 \leq \theta \leq \pi$

$\int_0^{\pi} \int_0^1 (\theta e^{\rho^2\sin\varphi})\rho\sin\varphi\, d\rho d\theta$

$u = \rho^2\sin\varphi; \ du = 2\rho\sin\varphi$

$= \int_0^{\pi} \frac{\theta e^{\rho^2\sin\varphi}}{2}\,\Big|_0^1 \, d\theta$

$= \int_0^{\pi} \frac{\theta(e^{\sin\varphi}-1)}{2}\, d\theta = \frac{\theta^2(e^{\sin\varphi}-1)}{4}\,\Big|_0^{\pi}$

$= \frac{\pi^2(e^{\sin\varphi}-1)}{4}$

Chapter 8

8.1: $f(x,z) = 3z^3x^2$; $\quad 0 \le x \le 1; \; z = 2$ $\int_0^1 3z^3x^2 \, dx = z^3x^3\big\|_0^1 = z^3 = 8.$	**8.2:** $f(x,y) = x \sin xy$; $\quad 0 \le y \le \pi; \; x = 1/2$ $\int_0^\pi x \sin xy \, dy = -\cos xy \big\|_0^\pi$ $= -\cos\left(\frac{1}{2} \cdot \pi\right) - [-\cos 0] = 1.$
8.3: $f(r,z) = 2r \sin rz$; $\quad 0 \le z \le 1; \; r = \pi$ $\int_0^1 2r \sin rz \, dz = -2\cos rz \big\|_0^1$ $= -2\cos(1 \cdot \pi) - [-2] = 4.$	**8.4:** $f(r,\theta) = 2 \cos r\theta$; $\quad 0 \le \theta \le 1/2; \; r = \pi$ $\int_0^{1/2} 2 \cos r\theta \, rd\theta = 2 \sin rz \big\|_0^{1/2}$ $= 2 \sin\left(\frac{1}{2} \cdot \pi\right) - [0] = 2.$
8.5: $f(\rho,\varphi) = \rho^2 e^{\cos \varphi}$; $\quad 0 \le \rho \le 1; \; \varphi = 0$ $\int_0^1 \rho^2 e^{\cos \varphi} \, d\rho = \frac{\rho^3}{3} e^{\cos \varphi}\big\|_0^1$ $= \frac{1}{3}e^1 = \frac{e}{3}.$	**8.6:** $f(\rho,\varphi) = \sin \varphi$; $\quad 0 \le \varphi \le \pi; \; \rho = 10$ $\int_0^\pi \sin \varphi \, \rho d\varphi = -\rho \cos \varphi \big\|_0^\pi$ $= -10[-1-1] = 20.$
8.7: A cylinder of radius R and height H $\int_0^R \int_0^{2\pi} \int_0^H rdzd\theta dr$ $= \int_0^R \int_0^{2\pi} Hrd\theta dr$ $= \int_0^R 2\pi H \, rdr = \pi HR^2$	**8.8:** $f(r,\theta,z) = r\theta z$ Over a cylinder of radius 2, height 3 $\int_0^2 \int_0^{2\pi} \int_0^3 r\theta z \, rdzd\theta dr$ $= \int_0^2 \int_0^{2\pi} \frac{9}{2}\theta r^2 \, d\theta dr$ $= \int_0^2 9\pi^2 r^2 dr = 24\pi^2$

8.9: $f(r, \theta, z) = \cos r\theta$
Over a cylinder
of radius 1/2, height 4

$$\int_0^{1/2} \int_0^{2\pi} \int_0^4 \cos r\theta \, r \, dz \, d\theta \, dr$$

$$= \int_0^{1/2} \int_0^{2\pi} 4 \cos r\theta \, r \, d\theta \, dr$$

$$= \int_0^{1/2} -4 \sin r 2\pi \, dr$$

$$= \frac{-2}{\pi} [\cos \pi - \cos 0] = \frac{4}{\pi}$$

8.10: $f(r, \theta, z) = ze^{r\theta}$
Over a cylinder
of radius 1, height 2

$$\int_0^1 \int_0^{2\pi} \int_0^2 ze^{r\theta} r \, dz \, d\theta \, dr$$

$$= \int_0^{1/2} \int_0^{2\pi} 2e^{r\theta} r \, d\theta \, dr$$

$$= \int_0^{1/2} 2[e^{2\pi r} - 1] \, dr$$

$$= \frac{1}{\pi}[e^\pi - 1] - \frac{1}{4}$$

8.11: $f(\rho, \varphi, \theta) = \rho^3$
Over a sphere of radius 2

$$\int_0^2 \int_0^\pi \int_0^{2\pi} \rho^3 \rho^2 \sin \varphi \, d\theta \, d\varphi \, d\rho$$

$$= \int_0^2 \int_0^\pi 2\pi\rho^5 \sin \varphi \, d\varphi \, d\rho$$

$$= \int_0^2 -2\pi(\cos \pi - \cos 0)\rho^5 \, d\rho$$

$$= \int_0^2 4\pi\rho^5 \, d\rho = \frac{2\pi}{3} 2^6 = \frac{2^7 \pi}{3}$$

8.12: $f(\rho, \varphi, \theta) = \cos^2 \varphi$
Over a sphere of radius 3

$$\int_0^3 \int_0^\pi \int_0^{2\pi} \cos^2 \varphi \, \rho^2 \sin \varphi \, d\theta \, d\varphi \, d\rho$$

$$= \int_0^3 \int_0^\pi 2\pi \cos^2 \varphi \rho^2 \sin \varphi \, d\varphi \, d\rho$$

$$= \int_0^3 \frac{2\pi}{3} [\cos^3 \pi - \cos^3 0] \rho^2 \, d\rho$$

$$= \int_0^3 \frac{4\pi}{3} \rho^2 \, d\rho = 12\pi$$

8.13: $f(\rho, \varphi, \theta) = \rho e^{\cos \varphi}$
Over a sphere of radius 1

$$\int_0^1 \int_0^\pi \int_0^{2\pi} \rho^3 e^{\cos \varphi} \sin \varphi \, d\theta \, d\varphi \, d\rho$$

$$= \int_0^1 \int_0^\pi 2\pi\rho^3 e^{\cos \varphi} \sin \varphi \, d\varphi \, d\rho$$

$$= \int_0^1 -2\pi\rho^3 [e^{-1} - e^1] \, d\rho$$

$$= \frac{2\pi}{4}\left[e - \frac{1}{e}\right] 1^4 = \frac{\pi}{2}\left[e - \frac{1}{e}\right]$$

8.14: $f(\rho, \varphi, \theta) = \frac{\ln(\cos \varphi)}{\rho^2}$
Over a sphere of radius 2

$$\int_0^2 \int_0^\pi \int_0^{2\pi} \frac{\ln(\cos \varphi)}{\rho^2} \rho^2 \sin \varphi \, d\theta \, d\varphi \, d\rho$$

$$= \int_0^2 \int_0^\pi 2\pi \ln(\cos \varphi) \sin \varphi \, d\varphi \, d\rho$$

$$= \int_0^2 2\pi \left[\frac{1}{\cos \pi} - \frac{1}{\cos 0}\right] d\rho$$

$$= \int_0^2 4\pi \, d\rho = 8\pi$$

Chapter 9

9.1: $\lim\limits_{x \to 1} \dfrac{x^2 - 2x + 1}{x^3 + 2x - 3}$ Form is $^0/_0$

$$= \lim\limits_{x \to 1} \frac{2x - 2}{3x^2 + 2} = \frac{0}{5} = 0$$

9.2: $\lim\limits_{x \to -2} \dfrac{x^2 - 4}{x^3 - 4x}$ Form is $^0/_0$

$$= \lim\limits_{x \to -2} \frac{2x}{3x^2 - 4} = \frac{-4}{8} = -\frac{1}{2}$$

9.3: $\lim\limits_{x \to 0} \dfrac{10x}{\sin 2x}$ Form is $^0/_0$

$$= \lim\limits_{x \to 0} \frac{10}{2 \cos 2x} = \frac{10}{2} = 5$$

9.4: $\lim\limits_{x \to \infty} \dfrac{5x^2 + 2}{\ln 2x}$ Form is $^\infty/_\infty$

$$= \lim\limits_{x \to \infty} \frac{10x}{2(\frac{1}{2x})} = 10(\infty^2) = \infty$$

9.5: $\lim\limits_{x \to \infty} \dfrac{x^4 + 3x^3 - 2x^2 + x - 1}{2x^4 + 5x^2 - 2}$ $^\infty/_\infty$

$$= \lim\limits_{x \to \infty} \frac{4x^3 + 9x^2 - 4x + 1}{8x^3 + 10x} \qquad ^\infty/_\infty$$

$$= \lim\limits_{x \to \infty} \frac{12x^2 + 18x - 4}{24x^2 + 10} \qquad ^\infty/_\infty$$

$$= \lim\limits_{x \to \infty} \frac{24x + 18}{48x} \qquad ^\infty/_\infty$$

$$= \lim\limits_{x \to \infty} \frac{24}{48} = \frac{1}{2}$$

9.6: $\lim\limits_{x \to \infty} x^2 4^{-x}$ Form is $\infty \cdot 0$

$$= \lim\limits_{x \to \infty} \frac{x^2}{4^x} \qquad ^\infty/_\infty$$

$$= \lim\limits_{x \to \infty} \frac{2x}{4^x \ln 4} \qquad ^\infty/_\infty$$

$$= \lim\limits_{x \to \infty} \frac{2}{4^x (\ln 4)^2} = \frac{2}{\infty} = 0$$

9.7: $\lim\limits_{x \to \infty} x \sin\dfrac{1}{x}$ Form is $\infty \cdot 0$

$$= \lim\limits_{x \to \infty} \frac{\sin x^{-1}}{x^{-1}} \qquad \text{Form is } ^0/_0$$

$$= \lim\limits_{x \to \infty} \frac{-x^{-2} \cos x^{-1}}{-x^{-2}} = \cos\frac{1}{\infty} = 1$$

9.8: $\lim\limits_{x \to 0} x \sin x$ Form is $0 \cdot 0$

The isn't one of the indeterminate forms. You can evaluate directly:

$$= \lim\limits_{x \to 0} x \sin x = 0 \cdot 0 = 0$$

9.9: $\lim_{x\to\infty} e^{-x}\ln x$ Form is $0\cdot\infty$

$$= \lim_{x\to\infty} \frac{\ln x}{e^x} \qquad \text{Form is } {}^{\infty}/_{\infty}$$

$$= \lim_{x\to\infty} \frac{{}^1/_x}{e^x} = \frac{0}{\infty} = 0$$

9.10: $\lim_{x\to 0} x^x$ Form is 0^0

$$L_{new} = \lim_{x\to 0} x\ln x \qquad \text{Form is } 0\cdot\infty$$

$$L_{new} = \lim_{x\to 0} \frac{\ln x}{x^{-1}} = \lim_{x\to 0} \frac{x^{-1}}{-x^{-2}} = 0$$

$$\lim_{x\to 0} x^x = L = e^{L_{new}} = e^0 = 1$$

9.11: $\lim_{x\to 0} (e^x)^{\cot x}$ Form is 1^{∞}

$$L_{new} = \lim_{x\to 0} \cot x \ln e^x$$

$$L_{new} = \lim_{x\to 0} x\cot x \qquad \text{Form is } 0\cdot\infty$$

$$L_{new} = \lim_{x\to 0} \frac{x}{\tan x} = \lim_{x\to 0} \frac{1}{\sec^2 x} = 1$$

$$L = e^{L_{new}} = e^1 = e$$

9.12: $\lim_{x\to 0} (\cos x)^{1/x}$ Form is 1^{∞}

$$L_{new} = \lim_{x\to 0} \frac{\ln\cos x}{x} \qquad \text{Form is } {}^0/_0$$

$$L_{new} = \lim_{x\to 0} \frac{-\sin x/\cos x}{1} = 0$$

$$L = e^{L_{new}} = e^0 = 1$$

9.13: $\lim_{x\to\infty} (x)^{e^{-x}}$ Form is ∞^0

$$L_{new} = \lim_{x\to\infty} e^{-x}\ln x$$

$$= \lim_{x\to\infty} \frac{\ln x}{e^x} = \lim_{x\to\infty} \frac{1}{xe^x} = \frac{1}{\infty} = 0$$

$$L = e^{L_{new}} = e^0 = 1$$

9.14: $\lim_{x\to 1} \left(\frac{1}{x-1}\right)^{\ln x}$ Form is ∞^0

$$L_{new} = \lim_{x\to 1} \left(\frac{1}{x-1}\right)^{\ln x}$$

$$= \lim_{x\to 1} \ln x \ln\left(\frac{1}{x-1}\right)$$

$$= \lim_{x\to 1} \frac{\ln\left(\frac{1}{x-1}\right)}{{}^1/_{\ln x}} = \lim_{x\to 1} \frac{x-1}{\frac{0-1/x}{(\ln x)^2}}$$

$$= \lim_{x\to 1} x(x-1)(\ln x)^2 = 0$$

$$L = e^{L_{new}} = e^0 = 1$$

Chapter 10

10.1: $a_1 = 4;\ a_k = 4 - a_{k-1}$					**10.2:** $a_1 = 5;\ a_k = a_{k-1} + 3$				
n	1	2	3	4	n	1	2	3	4
a_k	4	0	-4	-8	a_k	5	8	11	14
a_n	-4(1-2)	-4(2-2)	-4(3-2)	-4(4-2)	a_n	3(1)+2	3(2)+2	3(3)+2	3(4)+2

10.1 (continued):

Sequence is $\{-4(n - 2)\}$

$$\lim_{n\to\infty} -4(n - 2) = \infty;\ \text{Diverges}$$

10.2 (continued):

Sequence is $\{3n + 2\}$

$$\lim_{n\to\infty} -3n + 2 = \infty;\ \text{Diverges}$$

10.3: $a_1 = 10;\ a_k = 10a_{k-1}$					**10.4:** $a_1 = 64;\ a_k = \frac{a_{k-1}}{2}$				
n	1	2	3	4	n	1	2	3	4
a_k	10	100	1000	10000	a_k	64	32	16	8
a_n	10^1	10^2	10^3	10^4	a_n	$128/2^1$	$128/2^2$	$128/2^3$	$128/2^4$

10.3 (continued):

Sequence is $\{10^n\}$

Geometric Sequence w/ $|r| = 10$; Diverges

10.4 (continued):

Sequence is $\left\{128 * \left(\frac{1}{2}\right)^n\right\}$

Constant (128) times a Geometric Sequence ($|r| = \frac{1}{2}$); Converges to 0

10.5: $\left\{\frac{5n^2+2}{\ln 2n}\right\}$

$$\lim_{n\to\infty} \frac{5n^2+2}{\ln 2n} = \lim_{n\to\infty} \frac{10n}{n^{-1}} = \infty$$

Diverges

10.6: $\{e^{-n}\ln n\}$

$$\lim_{n\to\infty} \frac{\ln n}{e^n} = \lim_{n\to\infty} \frac{1/n}{e^n} = \frac{1/\infty}{\infty} = 0$$

Converges to 0

10.7: $\left\{\left(\frac{1}{n}\right)^n\right\}$

$$\lim_{n\to\infty} \left(\frac{1}{n}\right)^n = 0^\infty = 0$$

Converges to 0

10.8: $\{n^{1/n}\}$

$$\lim_{n\to\infty} (n)^{1/n} = \infty^0$$

$$L_{new} = \lim_{n\to\infty} \frac{\ln n}{n} = \lim_{n\to\infty} \frac{1/n}{n} = 0$$

$$L = e^{L_{new}} = 1;\ \text{Converges to 1}$$

10.9: $\left\{2 - \frac{1}{n}\right\}$	**10.10:** $\left\{\left(1 + \frac{2}{n}\right)^{2n}\right\}$
$\lim\limits_{n\to\infty} 2 - \frac{1}{n} = 2$ Converges to 2 Alternatively, you could point to the fact that as n increases, a_n increases, and with an upper bound of 2, $\therefore$ it's a bounded, monotonic, infinite sequence.	$\lim\limits_{n\to\infty} \left(1 + \frac{2}{n}\right)^{2n} = 1^\infty \to$ $L_{new} = \lim\limits_{n\to\infty} 2n \ln\left(1 + \frac{2}{n}\right)$ $\lim\limits_{n\to\infty} \frac{-2n^{-2}/(1+2n^{-1})}{-n^{-2}/2} = \lim\limits_{n\to\infty} \frac{n}{n+2}$ $\lim\limits_{n\to\infty} \frac{1}{1} = 1; L = e^{L_{new}} = e$ Converges to e
10.11: $\left\{2 + \frac{1}{2^n}\right\}$	**10.12:** $\left\{(-1)^{n+1}10^{-n}\right\}$
$\lim\limits_{n\to\infty} 2 + \frac{1}{2^n} = 2 + \lim\limits_{n\to\infty} \left(\frac{1}{2}\right)^n$ The latter term is a geometric sequence with $\|r\| = \frac{1}{2}$ which converges to 0, so the whole sequence converges to $2 + 0 = 2$	$\lim\limits_{n\to\infty} (-1)^{n+1}10^{-n}$ $= \lim\limits_{n\to\infty} -(-1)^n 10^{-n}$ $= \lim\limits_{n\to\infty} \frac{-(-1)^n}{10^n} = \lim\limits_{n\to\infty} -\left(\frac{-1}{10}\right)^n$ Geometric sequence with $\|r\| = \frac{1}{10}$; Converges to 0
10.13: $\left\{4^{-n}\cos 4n\right\}$	**10.14:** $\left\{\frac{n^3+3}{n^2+4n-2}\right\}$
$-1 \le \cos 4n \le 1$ $\lim\limits_{n\to\infty} \left(\frac{1}{4}\right)^n = 0$ $\frac{-1}{4^n} \le \frac{\cos 4n}{4^n} \le \frac{1}{4^n}$ $0 \le \frac{\cos 4n}{4^n} \le 0$; Converges to 0	$\lim\limits_{n\to\infty} \frac{n^3+3}{n^2+4n-2}$ $= \lim\limits_{n\to\infty} \frac{3n^2}{2n+4} = \lim\limits_{n\to\infty} \frac{6n}{2} = \infty;$ Diverges

Chapter 11

11.1: $\sum \frac{4n^2+2n+5}{3n^2-3n+5}$	**11.2:** $\sum \frac{n^3}{3n^3+2n^2-2n-3}$
$\lim\limits_{n\to\infty} a_n = \lim\limits_{n\to\infty} \frac{4n^2+2n+5}{3n^2-3n+5}$ $= \lim\limits_{n\to\infty} \frac{8n+2}{6n-3} = \lim\limits_{n\to\infty} \frac{8}{6} = \frac{4}{3} \neq 0$ <p align="center">Diverges</p>	$\lim\limits_{n\to\infty} \frac{n^3}{3n^3+2n^2-2n-3}$ $= \lim\limits_{n\to\infty} \frac{3n^2}{9n^2+4n-2}$ $= \lim\limits_{n\to\infty} \frac{6n}{18n+4} = \frac{6}{18} \neq 0$ <p align="center">Diverges</p>
11.3: $\sum \left(1+\frac{3}{n}\right)^n$	**11.4:** $\sum n^{1/2n}$
$\lim\limits_{n\to\infty} \left(1+\frac{3}{n}\right)^n \to \lim\limits_{n\to\infty} n \ln\left(1+\frac{3}{n}\right)$ $= \lim\limits_{n\to\infty} \frac{-3n^{-2}/(1+\frac{3}{n})}{-n^{-2}} = -3$ <p align="center">$L = e^{-3} \neq 0, \therefore$ Diverges</p>	<p align="center">$\lim\limits_{n\to\infty} n^{1/2n} \to$</p> $\lim\limits_{n\to\infty} \frac{1}{2n} \ln n = \lim\limits_{n\to\infty} \frac{1/n}{2} = 0$ <p align="center">$L = e^0 = 1 \neq 0, \therefore$ Diverges</p>
11.5: $2 + 0.2 + 0.02 + \cdots$ The pattern is : $\frac{2}{1} + \frac{2}{10} + \frac{2}{100} + \cdots + \frac{2}{10^{n-1}} + \cdots$ This is a geometric series with <p align="center">$a = 2, \quad r = \frac{1}{10}.$</p> By definition, it converges with sum: <p align="center">$\frac{a}{1-r} = 2.\overline{22}$</p>	**11.6:** $0.04 + 0.0004 + 0.000004 + \cdots$ The pattern is: $\frac{4}{100} + \frac{4}{100^2} + \frac{4}{100^3} + \cdots + \frac{4}{100^n} + \cdots$ Or, alternatively: $\frac{0.04}{1} + \frac{0.04}{100} + \frac{0.04}{100^3} + \cdots + \frac{.04}{100^{n-1}} + \cdots$ It's a convergent geometric series: <p align="center">$a = .04, \quad r = \frac{1}{100}$</p> with sum $\frac{a}{1-r} = 0.\overline{04}$

11.7: $\sum \left(\frac{1}{3}\right)^n \left(\frac{1}{4}\right)^{n-1}$

$$= \sum \frac{1}{3}\left(\frac{1}{3}\right)^{n-1}\left(\frac{1}{4}\right)^{n-1} = \frac{1}{3}\sum \left(\frac{1}{12}\right)^{n-1}$$

It's a convergent geometric series:
$$a = 1, \quad r = \frac{1}{12}$$

with sum $\frac{1}{3} \times \frac{a}{1-r} = 0.\overline{36}$

11.8: $3 + \frac{3}{4} + \frac{3}{9} + \frac{3}{16} + \frac{3}{25} + \cdots$

This is a p-series with form:
$$3\sum \frac{1}{n^2}$$

It does converge ($p > 1$), and the 5th, 10th, and 15th partial sums are:

$S_5 = 4.39; \ S_{10} = 4.65; \ S_{15} = 4.74$

as calculated by hand.

11.9: $5 + \frac{5}{2} + \frac{5}{3} + \frac{5}{4} + \cdots$

This series has the form:
$$5\sum \frac{1}{n}$$

This is a p-series with $p = 1$, also called a harmonic series. The series diverges.

11.10: $\sum \left[\frac{1}{2n-1} - \frac{1}{2n+1}\right]$

The first terms of this series are:

$$1 - \frac{1}{3} + \frac{1}{3} - \frac{1}{5} + \frac{1}{5} - \frac{1}{7} \cdots - \frac{1}{2n+1} \cdots$$

The nth Sum is:

$$S_n = 1 - \frac{1}{2n+1}.$$

The sum of the series is then:

$$S = \lim_{n \to \infty} 1 - \frac{1}{2n+1} = 1$$

therefore, it converges.

11.11: $\sum \left[\left(\frac{1}{2}\right)^n + \left(\frac{2}{n}\right)^n \right]$

The series can be split in two parts:

$$\sum \left(\frac{1}{2}\right)^n = \frac{1}{2} \sum \left(\frac{1}{2}\right)^{n-1}$$

is a convergent geometric series that sums to 1.

Convergence of $\sum \left(\frac{2}{n}\right)^n$ can be evaluated with the nth term test:

$$\lim_{n \to \infty} \left(\frac{2}{n}\right)^n \to \lim_{n \to \infty} n \ln \left(\frac{2}{n}\right)$$

$$= \lim_{n \to \infty} \frac{-2n^{-2}/(^n/_2)}{-n^{-2}} = 0;$$

$L = e^0 = 1 \neq 0 \therefore$ Diverges.

The sum of both series diverge.

11.12: $\sum \left[\left(\frac{1}{3}\right)^n + \left(\frac{1}{4}\right)^{n-1} \right]$

The series can be split in two parts:

$$\sum \left(\frac{1}{3}\right)^n = \frac{1}{3} \sum \left(\frac{1}{3}\right)^{n-1}$$

is a convergent geometric series that sums to $\frac{1}{3} \times \frac{1}{1-1/3} = \frac{1}{2}$.

$$\sum \left(\frac{1}{4}\right)^{n-1}$$

is a convergent geometric series that sums to $\frac{1}{1-1/4} = \frac{4}{3}$.

The final series converges with sum:

$$\frac{1}{2} + \frac{4}{3} = \frac{11}{6}.$$

11.13: $\sum \frac{3}{n(n+1)}$

The summation operator can be expressed as: $3 \times \left(\frac{A}{n} - \frac{B}{n+1}\right)$

From this we get: $An + A - Bn = 1$
$A - B = 0; A = 1 \to B = 1$

Plugging in, we find a converging telescoping series:

$$\sum \frac{3}{n(n+1)} = 3 \sum \left[\frac{1}{n} - \frac{1}{n+1}\right]$$

Following from the chapter example:

$$S = 3 \times \lim_{n \to \infty} \left(1 - \frac{1}{n+1}\right) = 3$$

11.14: $\sum \frac{1}{n^2+3n+2}$

The summation operator can be expressed as: $\left(\frac{A}{n+1} - \frac{B}{n+2}\right)$

From this we get:
$$An + 2A - Bn - B = 1$$
$$A - B = 0; 2A - B = 1 \to$$
$$A = B = 1$$

Plugging in: $\sum \left[\frac{1}{n+1} - \frac{1}{n+2}\right]$

$$\frac{1}{2} - \frac{1}{3} + \frac{1}{3} - \frac{1}{4} + \cdots - \frac{1}{n+2} + \cdots$$

$$S_n = \frac{1}{2} - \frac{1}{n+2}; \ S = \lim_{n \to \infty} S_n = \frac{1}{2}$$

Therefore the series converges.

Chapter 12

12.1: $\sum \dfrac{n^2}{n^3+3}$	**12.2:** $\sum \dfrac{n^2}{(n^3+3)^2}$
$\lim_{t\to\infty} \int_1^t \dfrac{x^2}{x^3+3}\,dx$	$\lim_{t\to\infty} \int_1^t \dfrac{x^2}{(x^3+3)^2}\,dx$
$= \frac{1}{3}\ln(t^3+3)\Big|_1^\infty = \infty$	$= \frac{-1}{3}(t^3+3)^{-1}\Big|_1^\infty = \frac{1}{12}$
Diverges by integration test	Converges by integration test
12.3: $\sum ne^{-n^2}$	**12.4:** $\sum \dfrac{\left|\cos\left(\frac{1}{n^2}\right)\right|}{n}$
$\lim_{t\to\infty} \int_1^t xe^{-x^2}\,dx = \frac{-1}{2}e^{-t^2}\Big|_1^\infty = \frac{1}{2e}$	
Converges by integration test	$\lim_{t\to\infty} \int_1^t \dfrac{\left|\cos\left(\frac{1}{x^2}\right)\right|}{x}\,dx$
	$= \frac{-1}{2}\sin\left(\frac{1}{t^2}\right)\Big|_1^\infty = \frac{\sin 1}{2}$
	Converges by integration test
12.5: $\sum \dfrac{1}{n2^n}$	**12.6:** $\sum \dfrac{\cos^2 n}{3^n}$
$2^n < n2^n \to \dfrac{1}{2^n} > \dfrac{1}{n2^n}$	$\cos^2 n \le 1 \to \dfrac{\cos^2 n}{3^n} \le \dfrac{1}{3^n}$
$\dfrac{1}{2^n} = \frac{1}{2}\left(\frac{1}{2}\right)^{n-1}$ is b_n of a convergent geometric series, so $\sum \dfrac{1}{n2^n}$ converges by basic comp. test	$\dfrac{1}{3^n} = \frac{1}{3}\left(\frac{1}{3}\right)^{n-1}$ is b_n of a convergent geometric series, so $\sum \dfrac{\cos^2 n}{3^n}$ converges by basic comp. test
12.7: $\sum \dfrac{n}{2n^2-1}$	**12.8:** $\sum \dfrac{2}{2+n^2}$
$2n^2 > 2n^2 - 1 \to \dfrac{n}{2n^2} < \dfrac{n}{2n^2-1}$	$n^2 < 2 + n^2 \to \dfrac{2}{n^2} > \dfrac{2}{2+n^2}$
$\dfrac{n}{2n^2} = \frac{1}{2}\left(\frac{1}{n}\right)$ is b_n of a divergent harmonic series, so $\sum \dfrac{n}{2n^2-1}$ diverges by basic comparison test	$\dfrac{2}{n^2} = 2\left(\frac{1}{n^2}\right)$ is b_n of a convergent p-series, so $\sum \dfrac{2}{2+n^2}$ converges by basic comparison test

12.9: $\sum \dfrac{n+1}{2n^2-1}$

$1/n$ is b_n of a diverging series.

$$\lim_{n\to\infty} \dfrac{n+1}{\frac{1}{n}(2n^2-1)}$$

$$= \lim_{n\to\infty} \dfrac{n^2+n}{(2n^2-1)} = \dfrac{1}{2} > 0$$

The series $\sum \dfrac{n+1}{2n^2-1}$ diverges by the limit comparison test

12.10: $\sum \dfrac{1}{\sqrt{n}+1}$

$\dfrac{1}{\sqrt{n}} = \left(\dfrac{1}{n}\right)^{1/2}$ is b_n of a diverging p-series.

$$\lim_{n\to\infty} \dfrac{\sqrt{n}}{\sqrt{n}+1} = \lim_{n\to\infty} \dfrac{n}{n+\sqrt{n}}$$

$$= \lim_{n\to\infty} \dfrac{1}{1+1/2\sqrt{n}} = 1 > 0$$

The series $\sum \dfrac{1}{\sqrt{n}+1}$ diverges by the limit comparison test

12.11: $\sum \dfrac{n^2+2}{n!}$

$$a_{n+1} = \dfrac{(n+1)^2+2}{(n+1)!}$$

$$\lim_{n\to\infty} \dfrac{a_{n+1}}{a_n} = \lim_{n\to\infty} \dfrac{n![(n+1)^2+2]}{(n+1)!(n^2+2)}$$

$$= \lim_{n\to\infty} \dfrac{[(n+1)^2+2]}{(n+1)(n^2+2)} = \overset{L'Hôpital's\ Rule}{\cdots} = 0 < 1$$

Series converges by the ratio test

12.12: $\sum \dfrac{n!}{ne^n}$

$$a_{n+1} = \dfrac{(n+1)!}{(n+1)e^{n+1}}$$

$$\lim_{n\to\infty} \dfrac{a_{n+1}}{a_n} = \lim_{n\to\infty} \dfrac{(n+1)!\,ne^n}{n!\,(n+1)e^{n+1}}$$

$$= \lim_{n\to\infty} \dfrac{(n+1)n}{(n+1)e} = \infty > 1$$

Series diverges by the ratio test

12.13: $\sum \dfrac{2^n}{n^2+2}$

$$a_{n+1} = \dfrac{2^{n+1}}{(n+1)^2+2}$$

$$\lim_{n\to\infty} \dfrac{a_{n+1}}{a_n} = \lim_{n\to\infty} \dfrac{2^{n+1}(n^2+2)}{2^n[(n+1)^2+2]}$$

$$= \lim_{n\to\infty} \dfrac{2(n^2+2)}{[(n+1)^2+2]} = \overset{L'Hôpital's\ Rule}{\cdots} = 2 > 1$$

Series diverges by the ratio test

12.14: $\sum \dfrac{2^n}{n^{n+2}}$

$$\lim_{n\to\infty} \sqrt[n]{a_n} = \lim_{n\to\infty} \left(\dfrac{2^n}{n^{n+2}}\right)^{1/n}$$

$$\lim_{n\to\infty} \dfrac{2}{n^{\left(1+\frac{2}{n}\right)}} = 0 < 1$$

Series converges by the root test

121

Chapter 13

13.1: $\sum(-1)^n \frac{n}{n^2+1}$	**13.2:** $\sum(-1)^n \frac{4}{2n^3-1}$

13.1: $\sum(-1)^n \frac{n}{n^2+1}$

n^{th} term test:

$$\lim_{n\to\infty} \frac{n}{n^2+1} = \lim_{n\to\infty} \frac{1}{2n} = 0$$

Limit Comparison test with divergent series $\sum\frac{1}{n}$:

$$\lim_{n\to\infty} \frac{n^2}{n^2+1} = \lim_{n\to\infty} \frac{2n}{2n} = 1 > 0;$$

$$\therefore \sum|a_n| \text{ diverges}$$

Alternating Series Test:

$$D_n\left[\frac{n}{n^2+1}\right] = \frac{n^2+1-2n^2-n}{(n^2+1)^2}$$

$$= \frac{-n^2-n+1}{(n^2+1)^2} < 0; \text{ Converges}$$

$$\therefore \text{ Conditionally Convergent}$$

13.2: $\sum(-1)^n \frac{4}{2n^3-1}$

n^{th} term test:

$$\lim_{n\to\infty} \frac{4}{2n^3-1} = 0$$

Limit Comparison test with convergent series $\sum\frac{1}{n^3}$:

$$\lim_{n\to\infty} \frac{4n^3}{2n^3-1} = \lim_{n\to\infty} \frac{24}{12} = 2 > 0;$$

$$\therefore \sum|a_n| \text{ converges}$$

$$\therefore \text{ Absolutely Convergent}$$

13.3: $\sum(-1)^n \frac{n}{n+1}$

n^{th} term test:

$$\lim_{n\to\infty} \frac{n}{n+1} = \lim_{n\to\infty} 1 \neq 0$$

$$\text{Divergent}$$

13.4: $\sum\left(\frac{n^3+1}{-n^2}\right)^n = \sum(-1)^n \left(\frac{n^3+1}{n^2}\right)^n$

n^{th} term test: $\lim_{n\to\infty} \left(\frac{n^3+1}{n^2}\right)^n$

$$= \lim_{n\to\infty} \left(\frac{n+1/n^2}{1}\right)^n = \infty^\infty \neq 0$$

$$\text{Divergent}$$

13.5: $\sum(-1)^n \left(\frac{4}{n}\right)^{1/n}$

n^{th} term test:

$$\lim_{n\to\infty} \left(\frac{4}{n}\right)^{1/n} \to \lim_{n\to\infty} \frac{\ln\left(\frac{4}{n}\right)}{n}$$

$$= \lim_{n\to\infty} \frac{n}{4}\left(\frac{4}{n^2}\right) = 0; L = e^0 = 1 \neq 0$$

Divergent

13.6: $\sum(-1)^n \frac{3}{1+\sqrt{n}}$

n^{th} term test: $\lim_{n\to\infty} \frac{3}{1+\sqrt{n}} = 0$

Limit Comparison test with divergent series $\sum \frac{1}{\sqrt{n}}$:

$$\lim_{n\to\infty} \frac{3\sqrt{n}}{1+\sqrt{n}} = \lim_{n\to\infty} \frac{3n}{\sqrt{n}+n}$$

$$= \lim_{n\to\infty} \frac{3}{1+1/(2\sqrt{n})} = 3 > 0$$

$$\therefore \sum|a_n| \text{ diverges}$$

Alternating Series Test:

$$D_n\left[\frac{3}{1+\sqrt{n}}\right] = \frac{-3(1+\sqrt{n})}{\left(1+\sqrt{n}\right)^2} < 0$$

$$\therefore \text{ converges}$$

Conditionally Convergent

13.7: $\sum(-1)^n \frac{n}{2^n}$

n^{th} term test:

$$\lim_{n\to\infty} \frac{n}{2^n} = \lim_{n\to\infty} \frac{1}{2^n \ln 2} = 0$$

Ratio Test with $a_{n+1} = \frac{n+1}{2^{n+1}}$:

$$\lim_{n\to\infty} \frac{(n+1)2^n}{n2^{n+1}} = \lim_{n\to\infty} \frac{n+1}{2n} = \frac{1}{2} < 1;$$

Absolutely Converges

13.8: $\sum(-1)^n \frac{e^n}{n!}$

Rather than attempt an nth term test, let's jump to an absolute convergence test and hope we get lucky.

Ratio Test with $a_{n+1} = \frac{e^{n+1}}{(n+1)!}$:

$$\lim_{n\to\infty} \frac{n! \, e^{n+1}}{(n+1)! e^n} = \lim_{n\to\infty} \frac{e}{n+1} = 0 < 1;$$

Absolutely Converges

13.9: $\sum(-1)^n \frac{n^2}{n^3+3}$

n^{th} term test:

$$\lim_{n\to\infty} \frac{n^2}{n^3+3} \to \lim_{n\to\infty} \frac{2}{6n} = 0$$

Integration Test:

$$\lim_{t\to\infty} \int_1^t \frac{x^2}{x^3+3}\,dx$$

$$= \frac{1}{3}\ln(t^3+3)\Big|_1^\infty = \infty;\ \text{diverges}$$

Alternating Series Test:

$$D_n\left[\frac{n^2}{n^3+3}\right] = \frac{-n^4+6n}{(n^3+3)^2} < 0$$

Since $n^4 \gg 6n$ for $n > 1$;
To deal with $n=1$, the series can be written as $\frac{1}{4} + \sum(-1)^n \frac{(n+1)^2}{(n+1)^3+3}$ which converges for $n \geq 1$; So, the series converges by "Latter Terms."

$\therefore$ Conditionally Convergent

13.10: $\sum(-1)^n \frac{1}{\sqrt{n+1}}$

n^{th} term test:

$$\lim_{n\to\infty} \frac{1}{\sqrt{n+1}} = 0$$

Integration Test:

$$\lim_{t\to\infty} \int_1^t (n+1)^{-1/2}\,dx$$

$$= 2(t+1)^{1/2}\Big|_1^\infty = \infty;\ \text{diverges}$$

Alternating Series Test:
$$D_n\left[(n+1)^{-1/2}\right]$$

$$= \frac{-1}{2}(n+1)^{-3/2} < 0;\ \text{converges}$$

$\therefore$ Conditionally Convergent

13.11: $\sum(-1)^n \frac{3^{1/n}}{n!}$

Skipping the nth term test, we apply the Ratio Test with $a_{n+1} = \frac{3^{1/(n+1)}}{(n+1)!}$:

$$\lim_{n\to\infty} \frac{n!\,3^{1/(n+1)}}{3^{1/n}(n+1)!} = \lim_{n\to\infty} \frac{3^{\left(\frac{1}{n+1}-\frac{1}{n}\right)}}{n+1}$$

$$= \lim_{n\to\infty} \frac{3^{-1/(n^2+n)}}{n+1} = \frac{1}{\infty} = 0 < 1;$$

Absolutely Converges

13.12: $\sum(-1)^n \frac{4^n}{n^{n+1}}$

Skipping the nth term test, we apply the Root Test:

$$\lim_{n\to\infty} \left(\frac{4^n}{n^{n+1}}\right)^{1/n}$$

$$= \lim_{n\to\infty} \frac{4}{n^{(1+1/n)}} = 0 < 1$$

Absolutely Converges

13.13: $\sum(-1)^n \dfrac{1}{n\sqrt{\ln(n)}}$

n^{th} term test:
$$\lim_{n\to\infty} \frac{1}{n\sqrt{\ln(n)}} = 0$$

Integration Test:
$$\lim_{t\to\infty} \int_1^t \frac{1}{n}(\ln(n))^{-1/2}dx$$

$$= 2(\ln(t))^{1/2}\big|_1^\infty = \infty; \text{ diverges}$$

Alternating Series Test:
$$D_n\left[(\ln(n))^{-1/2}\right]$$

$$= \frac{-1}{2}(\ln(n))^{-3/2} < 0 \text{ for } n > 1;$$

However, at $n = 1$, we have the first term as:

$$\frac{-1}{\sqrt{0}} = -\infty$$

So, if the first term is included, the series diverges as per "Properties of Addition"

Conditionally Convergent for $n > 1$
Divergent for $n \geq 1$

13.14: $\sum(-1)^n \dfrac{n^3+3n}{n^2+3n-2}$

n^{th} term test:
$$\lim_{n\to\infty} \frac{n^3+3n}{n^2+3n-2}$$

$$\lim_{n\to\infty} \frac{3n^2+3}{2n+3} = \lim_{n\to\infty} \frac{6n}{2} = \infty \neq 0$$

Diverges

Chapter 14

14.1: $\sum \frac{4x^n}{n^2}; n > 0$ $$\lim_{n \to \infty} \left| \frac{u_{n+1}}{u_n} \right| = \lim_{n \to \infty} \left| \frac{n^2 4 x^{n+1}}{(n+1)^2 4 x^n} \right|$$ $$\lim_{n \to \infty} \left| \frac{n^2 x}{(n+1)^2} \right| = \cdots = |x| < 1$$ At $x = \pm 1$, the series becomes $\sum (\pm 1)^n \frac{4}{n^2}$ which is an absolutely convergent p-series, so the interval is $|x| \leq 1$.	**14.2:** $\sum \frac{nx^n}{2^n}$ $$\lim_{n \to \infty} \left| \frac{u_{n+1}}{u_n} \right| = \lim_{n \to \infty} \left| \frac{2^n (n+1) x^{n+1}}{2^{n+1} n x^n} \right|$$ $$\lim_{n \to \infty} \left| \frac{(n+1)x}{2n} \right| = \cdots = \left| \frac{x}{2} \right| < 1$$ Converges for $|x| < 2$; At $x = \pm 2$, the series becomes $\sum (\pm 1)^n n$ which is a divergent series (by the nth term test), so the interval is $|x| < 2$.
14.3: $\sum \frac{x^n}{n!}$ $$\lim_{n \to \infty} \left| \frac{u_{n+1}}{u_n} \right| = \lim_{n \to \infty} \left| \frac{n! \, x^{n+1}}{(n+1)! \, x^n} \right|$$ $$= \lim_{n \to \infty} \left| \frac{x}{n+1} \right| = 0 < 1 \; \forall \, x;$$ Series converges for all x	**14.4:** $\sum \frac{3^{1/n} x^n}{n!}$ $$\lim_{n \to \infty} \left| \frac{u_{n+1}}{u_n} \right| = \lim_{n \to \infty} \left| \frac{n! \, 3^{1/(n+1)} \, x^{n+1}}{(n+1)! \, 3^{1/n} x^n} \right|$$ $$= \lim_{n \to \infty} \left| \frac{3^{1/(n+1)} x}{3^{1/n} (n+1)} \right| = 0 < 1 \; \forall \, x;$$ Series converges for all x
14.5: $\sum \frac{(4x)^n}{n^{n+1}}$ $$\lim_{n \to \infty} \left| \sqrt[n]{u_n} \right| = \lim_{n \to \infty} \left| \frac{4x}{n^{1+1/n}} \right|$$ $$= \lim_{n \to \infty} \left| \frac{4x}{n} \right| = 0 < 1 \; \forall \, x;$$ Series converges for all x	**14.6:** $\sum \frac{nx^n}{n+1}$ $$\lim_{n \to \infty} \left| \frac{u_{n+1}}{u_n} \right| = \lim_{n \to \infty} \left| \frac{(n+1)^2 x^{n+1}}{n(n+2) x^n} \right|$$ $$= \lim_{n \to \infty} \left| \frac{(n^2 + 2n + 2)x}{n^2 + 2n} \right| \to |x| < 1;$$ At $x = \pm 1$, the series becomes $\sum (\pm 1)^n \frac{n}{n+1}$ which is a divergent series (by the nth term test), so the interval is $|x| < 1$.

14.7: $\dfrac{2}{1-(x-1)^3}$

This is a sum of a geometric series with $a = 2; r = (x-1)^3$:

The series is then:

$$\sum 2[(x-1)^3]^n = \sum 2(x-1)^{3n}$$

14.8: $\dfrac{6(x-1)^2}{(1-(x-1)^3)^2}$

This is the derivative of $\dfrac{2}{1-(x-1)^3}$ so

$$\frac{6(x-1)^2}{(1-(x-1)^3)^2} = \sum_{n=0}^{\infty} D_x[2(x-1)^{3n}]$$

$$= \sum_{n=1}^{\infty} 6n\,(x-1)^{3n-1}$$

14.9: $\dfrac{1}{4+2x} = \dfrac{1/4}{1+x/2}$

This is a sum of a geometric series with $a = \dfrac{1}{4}; r = -x/2$:

The series is then:

$$\sum \frac{1}{4}\left(-\frac{x}{2}\right)^n = \sum \frac{(-1)^n}{4}\left(\frac{x}{2}\right)^n$$

14.10: $\dfrac{-2}{(4+2x)^2}$

This is the derivative of $\dfrac{1}{4+2x}$ so

$$\frac{-2}{(4+2x)^2} = \sum_{n=0}^{\infty} D_x\left[\frac{(-1)^n}{4}\left(\frac{x}{2}\right)^n\right]$$

$$= \sum_{n=1}^{\infty} \frac{n(-1)^n x^{n-1}}{4\cdot 2^n}$$

14.11: $\dfrac{1}{2}\ln(4+2x)$

This is the integral of $\dfrac{1}{4+2x}$ so

$$\frac{1}{2}\ln(4+2x) = \sum \int_0^x \frac{(-1)^n}{4}\left(\frac{t}{2}\right)^n dt$$

$$= \sum \frac{(-1)^n}{4(n+1)2^n} x^{n+1}$$

14.12: $\dfrac{x}{1-x^2}$

This is a sum of a geometric series with $a = x; r = x^2$:

The series is then:

$$\sum x(x^2)^n = \sum x^{2n+1}$$

14.13: $\dfrac{1+x^2}{(1-x^2)^2}$

This is the derivative of $\dfrac{x}{1-x^2}$ so

$$\frac{1+x^2}{(1-x^2)^2} = \sum_{n=0}^{\infty} D_x[x^{2n+1}]$$

$$= \sum_{n=1}^{\infty}(2n+1)x^{2n}$$

14.14: $-\dfrac{1}{2}\ln(1-x^2)$

This is the integral of $\dfrac{x}{1-x^2}$ so

$$-\frac{1}{2}\ln(1-x^2) = \sum \int_0^x t^{2n+1} dt$$

$$= \sum \frac{x^{2n+2}}{2n+2}$$

Chapter 15

15.1: $\cos x$	15.2: $\ln(1 + x)$
$$f(x) = \sum_{n=0}^{\infty} \frac{f^{(n)}(0)}{n!} x^n$$ $$f^{(0)}(0) = \cos(0) = 1;$$ $$f^{(1)}(0) = -\sin(0) = 0;$$ $$f^{(2)}(0) = -\cos(0) = -1;$$ $$f^{(3)}(0) = \sin(0) = 0; \cdots$$ $$f^{(2n)}(0) = (-1)^n; f^{(2n+1)}(0) = 0;$$ $$\cos x = \sum_{n=0}^{\infty} (-1)^n \frac{x^{2n}}{(2n)!}$$ $$= 1 - \frac{x^2}{2!} + \frac{x^4}{4!} - \frac{x^6}{6!} + \cdots$$ Note: Because $\sum \frac{x^n}{n!}$ converges for all x (see problem 14.3), the interval of convergence is $(-\infty, \infty)$	$$f^{(0)}(0) = \ln 1 = 0;$$ $$f^{(1)}(0) = \frac{1}{1+x}\Big|_0 = 1;$$ $$f^{(2)}(0) = \frac{-1}{(1+x)^2}\Big|_0 = -1;$$ $$f^{(3)}(0) = \frac{2}{(1+x)^3}\Big|_0 = 2!;$$ $$f^{(4)}(0) = \frac{-3 \cdot 2}{(1+x)^4}\Big|_0 = -3!; \cdots$$ $$f^{(n)}(0) = (-1)^{n-1}(n-1)!$$ Or, equivalently: $$f^{(n+1)}(0) = (-1)^n(n)!$$ $$\ln(1 + x) = \sum_{n=0}^{\infty} \frac{(-1)^n n!}{(n+1)!} x^{n+1}$$ $$\ln(1 + x) = \sum_{n=0}^{\infty} \frac{(-1)^n}{n+1} x^{n+1}$$ $$= x - \frac{x^2}{2} + \frac{x^3}{3} - \frac{x^6}{4} + \cdots$$ Note: The ratio test gives: $$\lim_{n \to \infty} \left|\frac{(n+1) x^{n+2}}{(n+2) x^{n+1}}\right| \to |x| < 1$$ and, since $\ln(1 + 1) = 0.6$ and $\ln(1 - 1) = \infty$, the interval of convergence is $(-1, +1]$

15.3: $\tan^{-1} x$

$$f^{(0)}(0) = \tan^{-1} 0 = 0;$$
$$f^{(1)}(0) = \frac{1}{1+x^2}\Big|_0 = 1;$$
$$f^{(2)}(0) = \frac{-2x}{(1+x^2)^2}\Big|_0 = 0;$$

$$f^{(3)}(0) = -2$$

$$\left(\frac{-2(1+x^2)^2 + 8x^2(1+x^2)}{(1+x^2)^4}\Big|_0\right);$$

After much bookkeeping:

$$f^{(4)}(0) = 0; \quad f^{(5)}(0) = 4!;$$

$$f^{(2n+1)}(0) = (-1)^n(2n)!$$

$$\tan^{-1} x = \sum_{n=0}^{\infty} \frac{(-1)^n(2n)!}{(2n+1)!} x^{2n+1}$$

$$= \sum_{n=0}^{\infty} \frac{(-1)^n}{2n+1} x^{2n+1}$$

$$= x - \frac{x^3}{3} + \frac{x^5}{5} - \frac{x^7}{7} + \cdots$$

Note: The ratio test gives:

$$\lim_{n\to\infty} \left|\frac{(2n+1)\, x^{2n+2}}{(2n+2)\, x^{2n+1}}\right| \to |x| < 1$$

and, since $\tan^{-1}(\pm 1) = \pm 0.79$, the interval of convergence is
$$[-1, +1]$$

15.4: $\sinh x$

$$f^{(0)}(0) = \sinh 0 = 0;$$
$$f^{(1)}(0) = \cosh 0 = 1;$$
$$f^{(2)}(0) = \sinh 0 = 0;$$
$$f^{(3)}(0) = \cosh 0 = 1;$$

$$f^{(2n+1)}(0) = 1$$

$$\sinh x = \sum_{n=0}^{\infty} \frac{1}{(2n+1)!} x^{2n+1}$$

$$= x + \frac{x^3}{3!} + \frac{x^5}{5!} + \frac{x^7}{7!} + \cdots$$

Note: Because $\sum \frac{x^n}{n!}$ converges for all x (see problem 14.3), the interval of convergence is $(-\infty, \infty)$

15.5: $\cosh x$

$$f^{(0)}(0) = \cosh 0 = 1;$$
$$f^{(1)}(0) = \sinh 0 = 0;$$
$$f^{(2)}(0) = \cosh 0 = 1;$$
$$f^{(3)}(0) = \sinh 0 = 0;$$

$$f^{(2n)}(0) = 1$$

$$\cosh x = \sum_{n=0}^{\infty} \frac{1}{(2n)!} x^{2n}$$

$$= 1 + \frac{x^2}{2!} + \frac{x^4}{4!} + \frac{x^6}{6!} + \cdots$$

Note: Because $\sum \frac{x^n}{n!}$ converges for all x (see problem 14.3), the interval of convergence is $(-\infty, \infty)$

15.6: e^{-x^2}

Since

$$e^x = 1 + x + \frac{x^2}{2!} + \frac{x^3}{3!} + \frac{x^4}{4!} + \cdots$$

for $x \in (-\infty, \infty)$

Then

$$e^{-x^2} = 1 - x^2 + \frac{x^4}{2!} - \frac{x^6}{3!} + \frac{x^8}{4!} + \cdots$$

for $-x^2 \in (-\infty, \infty)$
or just $x \in (-\infty, \infty)$

15.7: $\ln 1.1$

Since

$$\ln(1 + x) = x - \frac{x^2}{2} + \frac{x^3}{3} - \frac{x^6}{4} + \cdots$$

Then

$$\ln(1 + 0.1) = 0.1 - \frac{10^{-2}}{2} + \frac{10^{-3}}{3}$$
$$- \frac{10^{-4}}{4} \cdots \cong 0.095308$$

Compare to the calculator result:

$$\ln(1.1) = 0.095310$$

15.8: $\tan^{-1} 0.4$

Since

$$\tan^{-1} x = x - \frac{x^3}{3} + \frac{x^5}{5} - \frac{x^7}{7} + \cdots$$

Then

$$\tan^{-1} 0.4 = 0.4 - \frac{.064}{3} + \frac{0.010}{5}$$
$$- \frac{0.002}{7} \cdots \cong 0.3804$$

Compare to the calculator result:

$$\tan^{-1} 0.4 = 0.3805$$

15.9: $\int_0^{0.1} x \sin x \ dx$

Using the results from the text:

$$\int_0^{0.1} x \sin x \ dx$$

$$\cong \int_0^{0.1} x \left[x - \frac{x^3}{3!} + \frac{x^5}{5!} \right] dx$$

$$= \frac{x^3}{3} - \frac{x^5}{5 \cdot 3!} + \frac{x^7}{7*5!} \Big|_0^{0.1} = 3.33 \cdot 10^{-4}$$

15.10: $\int_0^{0.1} x^2 \tan^{-1} x \ dx$

Since $\tan^{-1} x \cong x - \frac{x^3}{3} + \frac{x^5}{5}$

$$\int_0^{0.1} x^2 \tan^{-1} x \ dx$$

$$\cong \int_0^{0.1} x^2 \left[x - \frac{x^3}{3} + \frac{x^5}{5} \right] dx$$

$$= \frac{x^4}{4} - \frac{x^6}{18} + \frac{x^8}{40} \Big|_0^{0.1} = 2.49 \cdot 10^{-5}$$

15.11: $\int_0^{0.1} x^2 \cos 2x \ dx$

Since $\cos x \cong 1 - \frac{x^2}{2!} + \frac{x^4}{4!}$, then

$$\cos 2x \cong 1 - \frac{4x^2}{2!} + \frac{16x^4}{4!}$$

$$\int_0^{0.1} x^2 \cos 2x \ dx$$

$$\cong \int_0^{0.1} x^2 - 2x^4 + \frac{16x^6}{24} dx$$

$$= \frac{x^3}{3} - \frac{2x^5}{5} + \frac{16x^7}{168} \Big|_0^{0.1} = 3.29 \cdot 10^{-4}$$

15.12: $\int_0^{0.1} x^3 \ln(1-x) \ dx$

Since $\ln(1-x) \cong -x - \frac{x^2}{2} - \frac{x^3}{3}$

$$\int_0^{0.1} x^3 \ln(1-x) \ dx$$

$$\cong \int_0^{0.1} -x^4 - \frac{x^5}{2} - \frac{x^6}{3} dx$$

$$= -\frac{x^5}{5} - \frac{x^6}{12} - \frac{x^7}{21} \Big|_0^{0.1}$$

$$= -2.09 \times 10^{-6}$$

15.13: $\int_0^{0.1} \sqrt{1 + \sqrt{x}} \ dx$

Using a binomial series:
$$(1 + \sqrt{x})^{1/2} \cong 1 + \frac{\sqrt{x}}{2} - \frac{x}{8}$$

$$\int_0^{0.1} \sqrt{1 + \sqrt{x}} \ dx$$

$$\cong \int_0^{0.1} 1 + \frac{\sqrt{x}}{2} - \frac{x}{8} dx$$

$$= x - \frac{1}{\sqrt{x}} - \frac{x^2}{16} \Big|_0^{0.1} = -3.063$$

15.14: $\int_0^{0.1} x \sqrt[4]{1 - x} \ dx$

Using a binomial series:
$$(1 - x)^{1/4} \cong 1 - \frac{x}{4} - \frac{3x^2}{32}$$

$$\int_0^{0.1} x \sqrt[4]{1 - x} \ dx$$

$$\cong \int_0^{0.1} x - \frac{x^2}{4} - \frac{3x^3}{32} dx$$

$$= \frac{x^2}{2} - \frac{x^3}{12} - \frac{3x^4}{128} \Big|_0^{0.1} = 4.92 \cdot 10^{-3}$$

Appendix D: Derivative Tables

Logarithms & Exponentials:

$$D_x[a^u] = a^u \ln a \, D_x[u]$$

$$D_x[e^u] = e^u D_x[u]$$

$$D_x[\log_a u] = \frac{1}{u \ln a} D_x[u]$$

$$D_x[\ln|u|] = \frac{1}{u} D_x[u]$$

Circular Trig Functions:

$$D_x[\sin(u)] = \cos(u) \, D_x[u]$$

$$D_x[\cos(u)] = -\sin(u) \, D_x[u]$$

$$D_x[\tan(u)] = \sec^{2(u)} D_x[u]$$

$$D_x[\sec(u)] = \sec(u)\tan(u) \, D_x[u]$$

$$D_x[\cot(u)] = -\csc^2(u) \, D_x[u$$

$$D_x[\csc(u)] = -\csc(u)\cot(u) \, D_x[u]$$

Inverse Circular Trig Functions:

$$D_x[\sin^{-1} u] = \frac{D_x[u]}{\sqrt{1-u^2}}$$

$$D_x[\cos^{-1} u] = \frac{-D_x[u]}{\sqrt{1-u^2}}$$

$$D_x[\tan^{-1} u] = \frac{D_x[u]}{1+u^2}$$

$$D_x[\cot^{-1} u] = \frac{-D_x[u]}{1+u^2}$$

$$D_x[\sec^{-1} u] = \frac{D_x[u]}{u\sqrt{u^2-1}}$$

$$D_x[\csc^{-1} u] = \frac{-D_x[u]}{u\sqrt{u^2-1}}$$

Hyperbolic Trig Functions:

$$D_x(\sinh u) = \cosh u \, D_x[u] \qquad\qquad D_x(\operatorname{csch} u) = -\operatorname{csch} u \coth u \, D_x[u]$$

$$D_x(\cosh u) = \sinh u \, D_x[u] \qquad\qquad D_x(\operatorname{sech} u) = -\operatorname{sech} u \tanh u \, D_x[u]$$

$$D_x(\tanh u) = \operatorname{sech}^2 u \, D_x[u] \qquad\qquad D_x(\coth u) = -\operatorname{csch}^2 u \, D_x[u]$$

Inverse Hyperbolic Trig Functions:

$$D_x[\sinh^{-1} u] = \frac{1}{\sqrt{u^2+1}} D_x[u] \qquad\qquad D_x[\operatorname{csch}^{-1} u] = \frac{-1}{u\sqrt{1+u^2}} D_x[u]$$

$$D_x[\cosh^{-1} u] = \frac{1}{\sqrt{u^2-1}} D_x[u] \qquad\qquad D_x[\operatorname{sech}^{-1} u] = \frac{-1}{u\sqrt{1-u^2}} D_x[u]$$

$$D_x[\tanh^{-1} u] = \frac{1}{1-u^2} D_x[u] \qquad\qquad D_x[\coth^{-1} u] = \frac{1}{1-u^2} D_x[u]$$

Appendix E: Integral Tables

Logarithms & Exponentials:

$$\int a^u du = \frac{a^u}{\ln a} + c$$

$$\int e^u du = e^u + c$$

$$\int a^u \ln a \, du = a^u + c$$

$$\int \frac{1}{u} du = \ln|u| + c$$

$$\int \log_a u \, du = u\left[\frac{\ln u}{\ln a} - \ln a\right] + c$$

$$\int \ln u \, du = u[\ln(u) - 1] + c$$

Circular Trig Functions:

$$\int \sin u \, du = -\cos u + c$$

$$\int \csc u \, du = \ln|\csc u - \cot u| + c$$

$$\int \cos u \, du = \sin u + c$$

$$\int \sec u \, du = \ln|\sec u + \tan u| + c$$

$$\int \tan u \, du = \ln|\sec u| + c$$

$$\int \cot u \, du = \ln|\sin u| + c$$

$$\int \sec^2 u \, du = \tan u + c$$

$$\int \sec u \tan u \, du = \sec u + c$$

$$\int \csc^2 u \, du = -\cot u + c$$

$$\int \csc u \cot u \, du = -\csc u + c$$

Inverse Circular Trig Functions:

$$\int \sin^{-1} u \, du = u \sin^{-1} u + \sqrt{1 - u^2} + c$$

$$\int \cos^{-1} u \, du = u \cos^{-1} u - \sqrt{1 - u^2} + c$$

$$\int \tan^{-1} u \, du = u \tan^{-1} u - \frac{1}{2}\ln|1 + u^2| + c$$

$$\int \csc^{-1} u \, du = u \csc^{-1} u \pm \ln(u + \sqrt{u^2 - 1}) + c \quad \begin{cases} +\to 0 < \csc^{-1} u < \pi/2 \\ -\to -\pi/2 < \csc^{-1} u < 0 \end{cases}$$

$$\int \sec^{-1} u \, du = u \sec^{-1} u \mp \ln(u + \sqrt{u^2 - 1}) + c \quad \begin{cases} -\to 0 < \sec^{-1} u < \pi/2 \\ +\to \pi/2 < \sec^{-1} u < \pi \end{cases}$$

$$\int \cot^{-1} u \, du = u \cot^{-1} u + \frac{1}{2}\ln|1 + u^2| + c$$

Hyperbolic Trig Functions:

$$\int \sinh u \, du = \cosh u + c$$

$$\int \cosh u \, du = \sinh u + c$$

$$\int \tanh u \, du = \ln|\cosh u| + c$$

$$\int \operatorname{sech}^2 u \, du = \tanh u + c$$

$$\int \operatorname{csch}^2 u \, du = -\coth u + c$$

$$\int \operatorname{csch} u \, du = \ln\left|\tanh\left(\frac{u}{2}\right)\right| + c$$

$$\int \operatorname{sech} u \, du = \tan^{-1}(\sinh u) + c$$

$$\int \coth u \, du = \ln|\sinh u| + c$$

$$\int \operatorname{sech} u \tanh u \, du = -\operatorname{sech} u + c$$

$$\int \operatorname{csch} u \coth u \, du = \operatorname{csch} u + c$$

Inverse Hyperbolic Trig Functions:

$$\int \sinh^{-1} u \, du = u \sinh^{-1} u - \sqrt{u^2 + 1} + c$$

$$\int \cosh^{-1} u \, du = u \cosh^{-1} u \mp \sqrt{u^2 - 1} + c \qquad \begin{cases} -\to \cosh^{-1} u > 0 \\ +\to \cosh^{-1} u < 0 \end{cases}$$

$$\int \tanh^{-1} u \, du = u \tanh^{-1} u + \frac{1}{2}\ln(1 - u^2) + c$$

$$\int \operatorname{csch}^{-1} u \, du = u \operatorname{csch}^{-1} u + \sinh^{-1} u + c$$

$$\int \operatorname{sech}^{-1} u \, du = u \operatorname{sech}^{-1} u + \sin^{-1} u + c$$

$$\int \coth^{-1} u \, du = u \coth^{-1} u + \frac{1}{2}\ln(u^2 - 1) + c$$

Forms with $u^2 + 1; u^2 - 1; 1 - u^2$:

$$\int \frac{du}{\sqrt{1-u^2}} = \sin^{-1} u + c$$

$$\int \frac{du}{1+u^2} = \tan^{-1} u + c$$

$$\int \frac{du}{u\sqrt{u^2-1}} = \sec^{-1} u + c$$

$$\int \frac{1}{\sqrt{u^2+1}} du = \sinh^{-1} u + c$$

$$\int \frac{1}{\sqrt{u^2-1}} du = \cosh^{-1} u + c$$

$$\int \frac{1}{1-u^2} du = \tanh^{-1} u + c$$

$$-\int \frac{du}{\sqrt{1-u^2}} = \cos^{-1} u + c$$

$$-\int \frac{du}{1+u^2} = \cot^{-1} u + c$$

$$-\int \frac{du}{u\sqrt{u^2-1}} = \csc^{-1} u + c$$

$$\int \frac{-1}{u\sqrt{1+u^2}} du = \operatorname{csch}^{-1} u + c$$

$$\int \frac{-1}{u\sqrt{1-u^2}} du = \operatorname{sech}^{-1} u + c$$

$$\int \frac{1}{1-u^2} du = \coth^{-1} u + c$$

Forms with $a^2 + u^2$; $a^2 - u^2$; $u^2 - a^2$:

$$\int \frac{du}{a^2+u^2} = \frac{1}{a}\tan^{-1}\frac{u}{a} + c \qquad \int \frac{u\,du}{a^2+u^2} = \frac{1}{2}\ln|a^2 + u^2| + c$$

$$\int \frac{du}{a^2-u^2} = \frac{1}{2a}\ln\left|\frac{u+a}{u-a}\right| + c \qquad \int \frac{u\,du}{a^2-u^2} = \frac{-1}{2}\ln|a^2 - u^2| + c$$

$$\int \frac{du}{u^2-a^2} = \frac{1}{2a}\ln\left|\frac{u-a}{u+a}\right| + c \qquad \int \frac{u\,du}{u^2-a^2} = \frac{1}{2}\ln|u^2 - a^2| + c$$

Forms with $\sqrt{a^2 + u^2}$:

$$\int \sqrt{a^2 + u^2}\,du = \frac{u}{2}\sqrt{a^2 + u^2} + \frac{a^2}{2}\ln\left|u + \sqrt{a^2 + u^2}\right| + c$$

$$\int \frac{\sqrt{a^2+u^2}}{u}\,du = \sqrt{a^2 + u^2} - a\ln\left|\frac{a+\sqrt{a^2+u^2}}{u}\right| + c$$

$$\int \frac{\sqrt{a^2+u^2}}{u^2}\,du = \frac{-\sqrt{a^2+u^2}}{u} + \ln\left|u + \sqrt{a^2 + u^2}\right| + c$$

$$\int \frac{du}{\sqrt{a^2+u^2}} = \ln\left|u + \sqrt{a^2 + u^2}\right| + c$$

$$\int \frac{du}{u\sqrt{a^2+u^2}} = \frac{-1}{a}\ln\left|\frac{a+\sqrt{a^2+u^2}}{u}\right| + c$$

$$\int \frac{du}{u^2\sqrt{a^2+u^2}} = \frac{-\sqrt{a^2+u^2}}{a^2 u} + c$$

Forms with $\sqrt{a^2 - u^2}$:

$$\int \sqrt{a^2 - u^2}\,du = \frac{u}{2}\sqrt{a^2 - u^2} + \frac{a^2}{2}\sin^{-1}\frac{u}{a} + c$$

$$\int \frac{\sqrt{a^2 - u^2}}{u}\,du = \sqrt{a^2 - u^2} - a\ln\left|\frac{a + \sqrt{a^2 - u^2}}{u}\right| + c$$

$$\int \frac{\sqrt{a^2 - u^2}}{u^2}\,du = \frac{-\sqrt{a^2 - u^2}}{u} - \sin^{-1}\frac{u}{a} + c$$

$$\int \frac{du}{\sqrt{a^2 - u^2}} = \sin^{-1}\frac{u}{a} + c$$

$$\int \frac{du}{u\sqrt{a^2 - u^2}} = \frac{-1}{a}\ln\left|\frac{a + \sqrt{a^2 - u^2}}{u}\right| + c$$

$$\int \frac{du}{u^2\sqrt{a^2 - u^2}} = \frac{-\sqrt{a^2 - u^2}}{a^2 u} + c$$

Forms with $\sqrt{u^2 - a^2}$:

$$\int \sqrt{u^2 - a^2}\,du = \frac{u}{2}\sqrt{u^2 - a^2} - \frac{a^2}{2}\ln\left|u + \sqrt{u^2 - a^2}\right| + c$$

$$\int \frac{\sqrt{u^2 - a^2}}{u}\,du = \sqrt{u^2 - a^2} - a\cos^{-1}\frac{a}{u} + c$$

$$\int \frac{\sqrt{u^2 - a^2}}{u^2}\,du = \frac{-\sqrt{u^2 - a^2}}{u} + \ln\left|u + \sqrt{u^2 - a^2}\right| + c$$

$$\int \frac{du}{\sqrt{u^2 - a^2}} = \ln\left|u + \sqrt{u^2 - a^2}\right| + c$$

$$\int \frac{du}{u\sqrt{u^2 - a^2}} = \frac{-1}{a}\sec^{-1}\frac{u}{a} + c$$

$$\int \frac{du}{u^2\sqrt{u^2 - a^2}} = \frac{\sqrt{u^2 - a^2}}{a^2 u} + c$$

Index

www.ingramcontent.com/pod-product-compliance
Lightning Source LLC
Chambersburg PA
CBHW051219200326
41519CB00025B/7171